Consensu

GENUS:
Gender in Modern Culture

12

Russell West-Pavlov (Berlin)
Jennifer Yee (Oxford)
Sabine Schülting (Berlin)

Consensuality
Didier Anzieu, gender
and the sense of touch

Naomi Segal

Amsterdam - New York, NY 2009

GENUS

GENDER IN MODERN CULTURE

This new series is a forum for exploring cultural articulations of gender relations in modern society. The series publishes searching and challenging work in current gender studies combining an interdisciplinary approach with a rigorous critique of various cultural media and their modes of production and consumption. Publications interrogate the cultural forms which articulate, legitimize, construct, contest or transform gender configurations in the modern age.

Cover illustrations by David Segal Hamilton, 2008.

The paper on which this book is printed meets the requirements of "ISO 9706:1994, Information and documentation - Paper for documents - Requirements for permanence".

ISBN: 978-90-420-2586-8
© Editions Rodopi B.V., Amsterdam - New York, NY 2009
Printed in the Netherlands

This book is dedicated to my mother Leah Segal.

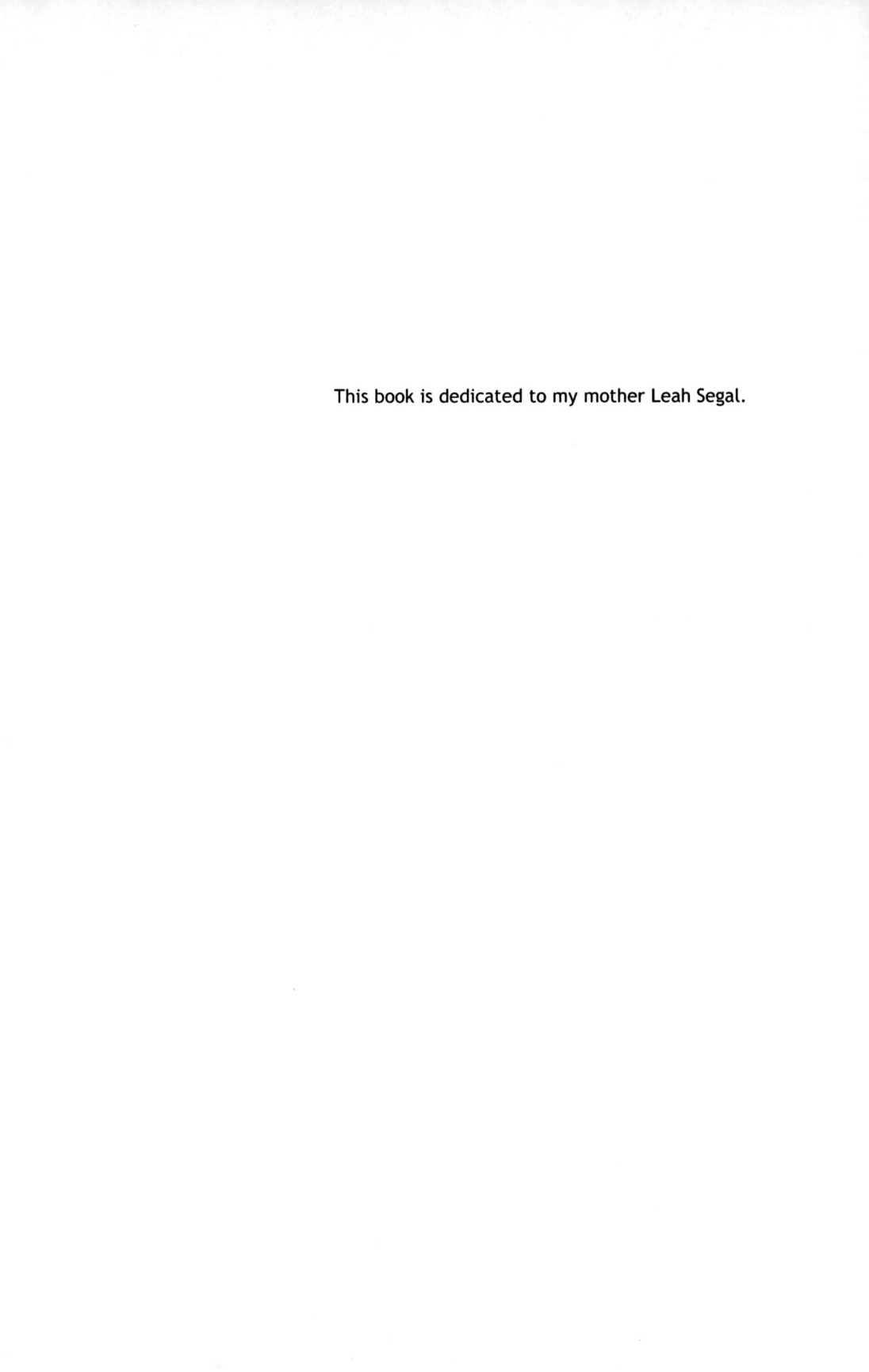

Acknowledgements

I am very grateful for permission to reprint sections of this book which have appeared elsewhere. Parts of chapter 5 are published with the permission of Verso Press and parts of chapter 7 with the permission of Cambridge Scholars Publishing. I am indebted to the Leverhulme Trust for the award of a Research Fellowship which enabled the work to begin in 2003. I should also like to thank colleagues and friends for all the discussions and suggestions that made this book possible, especially Christine Anzieu-Premmereur, Tom Baldwin, Chris Frith, John Forrester, Manucha Lisboa, Victoria Reid, Peter Turberfield and correspondents from jiscmails *francofil* and *german-studies*. Russ West-Pavlov has been a very genial editor, and I also owe a great debt to Richard Clark for his help with the knottier technical problems. Special thanks go to Rachel, David and Mat for the cover images.

Contents

Foreword	1
Chapter 1: Anzieu's life	7
Chapter 2: Anzieu's theory	31
Chapter 3: Anzieu and gender	55
Chapter 4: Gide's skin	79
Chapter 5: Diana's radiance	101
Chapter 6: The surface of things	119
Chapter 7: In the skin of the other	143
Chapter 8: Love	169
Chapter 9: Loss	201
Notes	241
Bibliography	249
Index	269

L'amour, tel qu'il existe dans la Société,
n'est que l'échange de deux fantaisies
et le contact de deux épidermes.
(Chamfort, 1796)

Love, as it exists in Society,
is nothing more than the exchange of two fantasies
and the contact of two epidermises.

Abbreviations

Throughout this book, all translations into English, unless otherwise attributed, are my own. In order to avoid the 'generic masculine' I have occasionally changed a French singular to an English plural, where no meaning is lost, and I generally use the pronoun 'they' in reference to ungendered singulars. With all translated quotations, reference is given to the original text. Further references to a cited text will appear after quotations; passages without page reference are from the last-cited page and page-numbers without specified text are similarly from the one last named. Unless otherwise stated, all italics are the authors'. In chapter 5, references without dates to journalistic articles on Princess Diana are all from 1997. Abbreviated titles used for texts cited in this book are given below in alphabetical order of abbreviation; full details can be found in the Bibliography.

AA	Annie Anzieu, *La femme sans qualité*
AA-a	Didier Anzieu, *L'Auto-analyse de Freud*
AABC	Didier Anzieu, *Mon ABC-daire*
AAc	Didier Anzieu et al., *L'activité de la pensée*
AAI	Didier Anzieu et al., *Autour de l'inceste*
ABe	Didier Anzieu, *Beckett*
ACD	Didier Anzieu, *Créer/Détruire*
ACI	Didier Anzieu, *Ces idées qui ont ébranlé la France*
ACO	Didier Anzieu, *Le Corps de l'œuvre*
ACon	Didier Anzieu, *Contes à rebours*
ACP	Didier Anzieu et al., *Les Contenants de pensée*

ADy	Didier Anzieu and Jacques-Yves Martin, *La dynamique des groupes restreints*
AEN	Didier Anzieu, *L'épiderme nomade et la peau psychique*
AEP	Didier Anzieu et al., *Les Enveloppes psychiques*
AGI	Didier Anzieu, *Le groupe et l'inconscient*
AMé	Didier Anzieu and Catherine Chabert, *Les méthodes projectives*
AMP	Didier Anzieu, *Le Moi-peau*
AOe	Didier Anzieu et al., *L'Œdipe : un complexe universel*
AP	Didier Anzieu, *Le Penser*
APA	Didier Anzieu, *Le Psychodrame analytique*
APL	Didier Anzieu et al., *Psychanalyse et langage*
APo	Didier Anzieu et al., *Portrait d'Anzieu avec groupe*
APP	Didier Anzieu, *Une peau pour les pensées*
APs	Didier Anzieu, *Psychanalyser*
AT	Nicolas Abraham and Maria Torok, *L'Écorce et le noyau*
CPD1	Maria Van Rysselberghe, *Cahiers de la petite dame* 1973
CPD2	Maria Van Rysselberghe, *Cahiers de la petite dame* 1974
CPD3	Maria Van Rysselberghe, *Cahiers de la petite dame* 1975
F-Ego	Sigmund Freud, *The Ego and the Id*
F-Int	Sigmund Freud, *The Interpretation of Dreams*
F-Moses	Sigmund Freud, *Moses & Monotheism*
F-MM	Sigmund Freud, 'Mourning and melancholia'
F-Rat	Sigmund Freud, 'Notes upon a case of obsessional neurosis (The "rat man")'
GJ1	André Gide, *Journal 1887-1925*
GJ2	André Gide, *Journal 1926-1950*
GJ3	André Gide, *Souvenirs et voyages*
GR	André Gide, *Romans*
HE	David Howes, *Empire of the Senses*
HV	David Howe, *Varieties of Sensory Experience*
ICC	Luce Irigaray, *Le corps-à-corps avec la mère*
ICS	Luce Irigaray, *Ce sexe qui n'en est pas un*
IED	Luce Irigaray, *Être deux*
IEDS	Luce Irigaray, *Éthique de la différence sexuelle*

IPE	Luce Irigaray, *Passions élémentaires*
IS	Luce Irigaray, *Speculum de l'autre femme*
JL	Jacques Lacan, *De la psychose paranoïaque dans ses rapports avec la personnalité*
LAQ	Emmanuel Lévinas, *Autrement qu'être*
LLB	Leconte de Lisle, *Poèmes barbares*
LLT	Leconte de Lisle, *Poèmes tragiques*
LTA	Emmanuel Lévinas, *Le temps et l'autre*
LTI	Emmanuel Lévinas, *Totalité et infini*
MPV	Merleau-Ponty, *Le Visible et l'invisible*
PSI	Alex Potts, *The Sculptural Imagination*
RB	V. S. Ramachandran and Sandra Blakeslee, *Phantoms in the Mind*
RHP	Élisabeth Roudinesco, *Histoire de la psychanalyse en France*, vol 2
RJL	Élisabeth Roudinesco, *Jacques Lacan*
RMLB	Rainer Maria Rilke, *Malte Laurids Brigge*
RNG	Rainer Maria Rilke, *Neue Gedichte*
RP	Rainer Maria Rilke, 'Puppen'
RR	Rainer Maria Rilke, 'Rodin'
SAC	Mahmoud Sami-Ali, *Corps réel, corps imaginaire*
SAE	Mahmoud Sami-Ali, *L'Espace imaginaire*
SEN	Jean-Paul Sartre, *L'Être et le néant*

Foreword

Most human beings have five senses, more or less. Everyday experience is "multisensual" (Rodaway 4) and "the senses are not merely passive receptors of particular kinds of environmental stimuli but are actively involved in the structuring of that information". I say more or less five, for the history and geography of the senses show that while that total is traditional, it is often disputed not only for the sake of precision but because of a general feeling that there must be something else.

> We have five senses in which we glory and which we recognise and celebrate, senses that constitute the sensible world for us. But there are other senses - secret senses, sixth senses, if you will - equally vital, but unrecognised, and unlauded. These senses, unconscious, automatic, had to be discovered. Historically, indeed, their discovery came late: what the Victorians vaguely called 'muscle sense' - the awareness of the relative position of trunk and limbs, derived from receptors in the joints and tendons - was only really defined (and named 'proprioception') in the 1890s. And the complex mechanisms and controls by which our bodies are properly aligned and balanced in space - these have only been defined in our own century and still hold many mysteries. (Sacks 1986: 68)

One suggestion lists ten basic senses, including four varieties of touch plus two of orientation (Rodaway 28, citing Gold 1980: 50). Others searching for the proverbial sixth sense cite extra-sensory perception (HV 258 and 290), desire (Serres 57-60), proprioception defined as "our totally intuitive sense of our own bodies" (Josipovici 110) or more rarefied abilities like that of the skilled wine-taster. Different cultures have more or fewer senses, or lay stress on different aspects. Of three non-literate societies cited by Constance Classen, "each has a very distinct way of making sense of the world: the Tzotzil accord primacy to heat in their cosmology, the Ongee to odor, and the Desana to color" (HE 148). Words for sensing are also variable, and often clustered: "the Hausa have one word for hearing, smelling, tasting and touching, understanding and emotional feeling" (Ritchie: HV 194); French too, of course, uses one word (*sentir*) for smelling and for both physical and emotional feeling.

However many senses we wish to number, it is interesting that, until recently, they were discussed only in order to be distinguished and separated. Since Aristotle, the senses have been placed in a hierarchical order, dependent either on proximity to the thing sensed or on the difference between human and animal. Thus "touch (and thereby taste) was found in all animals and so became the lowliest sense [... Aristotle] posed a hierarchical order of the senses, from most to least valuable: vision-hearing-smell-taste-touch" (Stewart: HE 61). Even if animals showed more skills than us with certain senses, theirs were intrinsically the inferior ones. This hierarchy slides into the other, for the last three of these are the "proximity" or "intimate senses" (Rodaway 26), devalued because they are deemed the furthest from thought, imagination and memory. As I have remarked elsewhere, these three senses are also the ones in which the nuances of active and passive perception are linguistically the least differentiated. If for sight and hearing we have three verbs:

> I look at the picture, I see the moon, I look tired,
> I listen to the music, I hear thunder, I sound interested,

for smell, taste, and touch, one verb has to stand in for all these functions:

> I smell a rose, I smell burning, I smell funny,
> I taste the soup, I taste a trace of cinnamon, it tastes bitter,
> I feel the velvet, I feel the sun on my face, I feel pretty.

But this could be a reason for suggesting that, far from being more blunt, the words we use for the proximate senses "do more work, convey more variation, carry more weight" (Segal 2001a: 18).

However undifferentiated language seems to think them, recent theory has turned back to these less favoured senses because, actually, they are better at imagining (Baudelaire), remembering (Proust) and of course loving.

Contemporary theory sees the senses as a multiplicity – hence the use of terms like "sensorium [...] sense ratio" (McLuhan: HE 43-51) or "sensotypes" (Wober: HV 33). To McLuhan sensing is a kaleidoscope (cited HV 167), to Serres a knot or an island (Serres 51-52), to Howes synaesthesia, the latter defined as a way of "short-circuiting the five sense model" (HE 292). It is the meeting of senses and sensations that most preoccupies current thinking: the "pluri-sensorial" (HV 6), "combinatory" (HV 167), "multidirectional [...] intersensoriality" (HE 12). And, as the rest of this book will explore, the multiplicity of the senses is most richly focused in the sense of touch (see Serres 82-84, Rodaway 28 and 44-54, Marks 2002: xiii, and Heller and Schiff 1-3).

Curiously, whichever way one looks at the lists of senses, touch is almost always found at one end.

> In the evolution of the senses the sense of touch was undoubtedly the first to come into being. Touch is the parent of our eyes, ears, nose, and mouth. It is the sense which became differentiated into the others, a fact that seems to be recognized in the age-old evaluation of touch as 'the mother of the senses'. (Montagu 3)

Among the three histories of feral children discussed by Constance Classen, Victor's faculties were ranked thus: "'The sense of smell is first and most perfected; taste is second, or rather these senses are but one; vision occupies the position of third importance, hearing the fourth, and touch the last'" (Pierre-Joseph Bonnaterre, cited HV: 49), whereas Kaspar Hauser "had an almost supernatural sense of touch. The touch of humans and animals gave him a sensation of heat or cold, at times so strong that he felt as if he had received a blow" (54). More generally, "the senses of *Homo sapiens* develop in a definite sequence, as (1) tactile, (2) auditory, and (3) visual. As the child approaches adolescence the order of precedence becomes reversed, as (1) visual, (2) auditory, and (3) tactile" (Montagu 314-15).

Indeed in infant development, of humans as well as animals, the stimulation of this sense is so crucial that "when the need for touch remains unsatisfied, abnormal behaviour will result" (Montagu 46) - "children need touch for survival" (Field 5).

For "the human sensorium [...] never exists in a natural state. Humans are social beings, and just as human nature itself is a product of culture, so is the human sensorium" (HE 3). In infants, the first version of this social interaction is the whole complex of holding, massage, breastfeeding understood as "reciprocal interstimulation" (Montagu 43) provided by the mother or primary caregiver. This is never only one-sided: among the Wolof of Senegal, "when a visitor arrives, male or female, often before any word is exchanged, he or she is handed a baby. This gesture is intended to 'mediate' the relation between adults" (HV 184). Touch is "a kind of communication between person and world, a corporeal situation rather than a cognitive positioning [...] Touch is direct and intimate, and perhaps the most truthful sense" (Rodaway 44); it is the sense we use to test the material reality of a thing by direct bodily perception (Josipovici 2 and 29). If, then, "the history of the senses has been, essentially, the history of their objectification" (Mazzio: HE 85), the "history of touch is, essentially, a history of resisting objectification" (86).

This book is not a history of the sense of touch but a more lateral exploration of the idea and meaning of that faculty. I begin from the principle that the body is not simply a lived object in space but an emissary: through its ability to touch it communicates something of the self to something of other selves. Of course we know little of the 'lives' of even our own internal organs (Leder, Jacques-Alain Miller), let alone the biographies of our feelings or the feelings of others. But we live this ignorance in a dramatic way and that drama is mediated through our organ of touch, the skin.

My first three chapters take a detailed look at the work of psychoanalyst Didier Anzieu, the major theorist of skin and containment, finishing with a chapter in which his writing is reexamined with gender questions in mind. In the other six chapters Anzieu may sometimes seem to disappear from view but his manner of thinking about the world and the way we live psychically as bodies among other bodies underlies the readings offered in them.

Chapters 4 and 5 take two individuals, André Gide and Princess Diana, and examine how their bodies function as subjects and objects of desire gendered masculine and feminine, in texts by and about them, in the way others think or talk of them. Chapter 6 looks at how poets and critics write about the desire of sculpted form and chapter 7 analyses a cinematic fantasy wherein one character uses another person's body to desire through. In the last two chapters, returning to Anzieu's theory alongside a collection of writers, film-makers, scientists and philosophers, I explore ways of understanding the idea of touch in relation to love and loss.

Thus my intention is to bring two kinds of material together: the theory of Didier Anzieu and his colleagues on the skin-ego, psychic envelopes, the dynamics of groups etc and a set of cultural objects, moments, angles or figures from the early nineteenth century to the present day. Before I can begin, I want to introduce the term I have taken for the title of this book.

The term 'consensuality' (*consensualité*) is used by Anzieu for the fifth function of the skin-ego, which brings together the perception of all the senses in one place – all our senses are sited somewhere on our skin – and thus stands for the coherence, coincidence or co-presence of perceptions (AMP 127; see also Serres 53, 59 and 83). Interestingly, Anzieu often also uses other terms: 'intersensoriality' or 'common sense', 'concordance', 'correspondence', 'convergence', 'contiguity' or 'consensus' (AMP 250, 253, 261, 264; and AP 15, 18, 35, 39, 56 59, 65, 104). The qualities of the sense of touch serve as "the ground or backcloth on which other sense qualities stand out as figures" (AP 35).

The word was borrowed from Donald Meltzer, who observes of a child's imaginative play: "we are envisaging attention as the strings which hold the senses together in consensuality" (Meltzer 1975: 12-13). It is from him that Anzieu takes the term for what can go wrong with this function: the consensual convergence may fall apart or, as Meltzer puts it, be passively "dismantled" by an autistic child when, "somewhat akin to allowing a brick wall to fall to pieces by the action of the weather, moss, fungi and insects" (12), the senses are made to attach singly and separately to the internal or external object that is "the most stimulating at that moment".

When consensuality works well and the skin-ego is in place, it can lead towards an ability to "seek out concordances and convergence" (AP 104) and forms of thought that emphasise analogy, the microcosm and "systems of correspondence between words, things and thoughts".

I have used this term as my title not only because, as these citations show, it is central to psychoanalytic thinking about the sense of touch, but also because this book is intended as an exercise in convergence, bringing together cultural processes, objects and theories so that they can work together consensually. For of course the most obvious meaning of the term - in both French and English - is to do with consensus and consent. Etymologically, as Anzieu notes, the idea of consensus derives from the "mutual agreement of the senses" (AP 35) and the idea of a 'common sense' brings the body together with the mind to create a basis for thought:

> The word 'sense' has a double meaning. 'Sense' means 'signification', which finds its most extensive organisation in language. 'The senses' are the body organs by which the human being makes contact with its surroundings and in the first place with its mother. The work by which the infant creates its psychic apparatus, and by which the patient and analyst create interpretation, consists of articulating 'sense' with 'the senses'. Words have value and bear meaning through their weight of flesh. The unconscious is not language; it is the body, [...] the intelligence of the body ('La psychanalyse encore', APs 268).

If the intelligence of the body is the basis of both sense and consent, the question of consensuality is also to do with human relations based on the sense of touch, most particularly the mother-child couple and the relation of desire. As Merleau-Ponty argues in *Le Visible et l'invisible*, every touch is double: it involves both an active and a passive experience; I feel myself feeling and felt. Thus, when I touch a sculpture or a piano, I am also dwelling on the surface of the body of the other. If pleasure is a tactical, tangential experience, is it also contagion? What are the limits of consent between bodies? Is love ever mutual?

Chapter 1: Anzieu's life

On 10 December 1890, a fatal accident befell a five-year-old child called Marguerite Pantaine, the daughter of a farming family in Chalvignac, in the *département* of Cantal in the Massif central. Here are two versions of the event:

> The family talk a lot about a violent emotion the mother suffered while she was pregnant with my patient. The eldest daughter died as the result of a tragic accident: she fell, before her mother's very eyes, into the wide-open door of a lighted stove and died very rapidly of severe burns. (JL 174-75)

> [My mother] was the third child in the family, the third or fourth... That's the problem. Before her, in fact, three daughters were born. The family lived in a large stone house close to the stable and the fields. The main room was heated by a large fireplace filled with big burning logs [...], and there were benches in it that you could sit on. This happened before my mother was born. It was a feast-day. Marguerite, the youngest of the three daughters, had an organdie dress on, ready to go to church. She'd been left for a moment in the charge of the eldest girl, the one who was to become my godmother. The child was lightly dressed, it was cold, she went up to the fire to warm herself... and was burnt alive. It was a dreadful shock for her parents and her two sisters. So my mother was conceived as a replacement for the dead child. And since she was another girl, they gave her the same name, Marguerite. The living dead, in a way... It's no coincidence that my mother spent her life finding ways to escape from the flames of hell... It was a way of accepting her fate, a tragic fate. My mother only spoke openly of this once. But I knew it as a family legend. I think her depression goes back to this untenable position. (APP 19-20)

The variation between these two versions shows how a "family legend" (which will also become a "fate" carried through several generations) twists and turns to serve the vagaries of later causalities. Was the dead girl the eldest or the youngest? Was it a gaping stove or an ingle-nook fireplace? Was it a sister or the mother who witnessed the terrible event and should have been taking better care? Was the mother pregnant when it happened or did she become so to replace her lost daughter, whether firstborn or lastborn? What they agree on is the shock to the whole family, and the long-lasting effects not just on the parents and living sisters but on the siblings that followed. In fact the child who died was the eldest, not the youngest, of the three sisters and the mother Jeanne was not pregnant with the second Marguerite when the first was killed, but (possibly) with a baby who was declared stillborn the following August (see Allouch; also Lessana 293-342). The second Marguerite was born in July 1892 (see Allouch 99-104, 156 and 248-60; also RJL 35).

The first version cited above is that of Jacques Lacan, who met Marguerite Pantaine, by then Marguerite Anzieu, in April 1931, when she was arrested and then sectioned for attacking a famous stage actress with a knife. Lacan wrote his doctoral thesis on Marguerite Anzieu under the pseudonym of 'Aimée', using her case-history as a prototype of the role of personality in psychopathic development. Paranoiac behaviour did not begin in the generation of 'Aimée': her mother Jeanne "has a reputation in the family for having 'persecution mania'. There is also an aunt (possibly Jeanne's sister) who quarrelled with everyone and had a reputation for rebellion and wild behaviour" (JL 174; see also Allouch 99). Jeanne Pantaine's reputation in the village as an "over-interpreter" (*interprétative*; see the use of *délire d'interprétation* as a synonym for paranoia) is, as Lacan describes it, as much an excessive sensitivity as an inclination to be difficult – "a vulnerability based on anxiety, which tended to switch quickly into suspicion" (JL 221) – and it is reasonable to suppose that it was exacerbated by the sudden death of her child. There was, in any case, a special closeness, a kind of friendship, between Jeanne and Marguerite (see JL 220 and Allouch 102), while it was a sister who took over maternal care of Marguerite and her three younger brothers.

This girl, who either was or was not meant to be keeping an eye on the first Marguerite when she fell into the fire (almost certainly not, since at three years old she was two years younger than her) was Élise Pantaine, known in the family as Eugénie or Nène. At fourteen she went to work for a paternal uncle, Guillaume, and married him in 1906 when he was twenty-eight and she was nineteen. Four years later, when, at eighteen, Marguerite started work in the administrative section of the Post Office, she went to live with them for a few months. She was then transferred to a village some way from her home and from there in 1913 to Melun, near Paris, where she met René Anzieu, also a Post Office employee, and they married in 1917. Meantime Élise, who could not have children after a total hysterectomy in 1914, came to live with the couple eight months after their wedding, for Guillaume had died as a result of war-wounds. When Marguerite went off the rails, she took over the care of the child and later also took the wife's place with René, though the family put up energetic opposition when he wished to get divorced and marry her after Marguerite's final discharge from hospital in 1943 (see Anzieu's account in RK 50).

The second version cited above is that of Didier Anzieu in a set of interviews conducted in 1983 when he was sixty. He goes on:

> I might put it this way – it sounds banal, but in my case it seems true: I became a psychoanalyst to care for my mother. Not so much to care for her in reality, even though I did succeed in helping her, in the last quarter of her life, to find a relatively happy, balanced life. What I mean is, to care for my mother in myself and other people. To care, in other people, for this threatening and threatened mother... (APP 20)

The double significance of the absent mother – threatening as much as threatened, damaged as much as damaging, and damaging, no doubt, because damaged – will remain throughout this complicated narrative.[1] Before we embark on the life and work of Didier Anzieu, we need to examine the complex configurations of similarity, engagement and replacement that preceded his birth.

To begin with, there is the doubling of given names. His mother was the fourth of her name within three generations: both her grandmothers were called Marguerite. While her dead sister was named only Marguerite, she was given her mother's name also, in second place: Marguerite Jeanne Pantaine, and, according to the *Petit Parisien* of 20 April 1931, she signed a book manuscript that she had submitted unsuccessfully to Flammarion with the first name of Jeanne (Allouch 192; see also 375-81). To enter and inhabit another person's name – in the Pantaine family everyone except Didier's mother seems to have had at least one nickname – is a sign of one position 'inside the skin of the other' that it will be the purpose of this book to examine.

Secondly, replacement in this family structure – a girl serving as mother to her younger sister, or even to her own disturbed mother, one child replacing another, a godmother replacing the birth-mother, who happens to be her sister, not only in relation to the son but also to the husband – all these replacements do not supersede the first relationship but coexist with it, as if 'ghosted' by it or wrapped inside it.

In interviews and his rare autobiographical writings, Anzieu speaks very little of his surrogate mother, referring to her, where he specifies at all who is included in the generic couple "my parents", as his godmother [*ma marraine*]. But his report that in family legend she was implicitly more at fault than the child's mother for the death of the first Marguerite suggests that Élise's role as carer in relation to the irresponsibility of either of the family 'madwomen' makes her the psychic substitute for the dangerous mother: neither mother nor dangerous, and therefore perhaps of little interest. What Didier becomes, as a 'good' analyst, is that figure of sanity that remains ghosted by the insanity it contains – 'contains' in both senses, as Anzieu's theory is devoted to exploring: that is, both 'enwraps' and 'holds in control'.

When did Marguerite Anzieu's madness begin? Anzieu has two, connected suggestions. The first is that it may have been set in train before she was born, with the burden of being the replacement child, the second that she was affected by her two pregnancies:

> Why was she depressed? Because of her sensitive character. It made her unable to deal with my birth, which brought back the terrible memory of the stillborn baby. And why this sensitivity? I think it was due to the circumstances of her own birth. [...]
> My mother only spoke openly of this once. But I knew it as a family legend. I think her depression goes back to this untenable position. She had put it off after the birth of her little girl who was stillborn, which seemed an implacable repetition of fate. And then my birth, which was successful, brought back that intolerable threat... (APP 19-20)

As described by her sister and brother, Marguerite was already as a child "very 'selfish'" [*personnelle*] (JL 220) and knew how to twist her normally tyrannical father round her little finger when it came to getting nicer hats or finer underwear. The family opposed her marriage, saying she had no domestic skills and was too inclined to go off into a dream; she agreed retrospectively that "I should have stayed with my mother" (220 and 241). But the first explicitly paranoid symptoms appeared when she became pregnant in 1921. The baby girl was healthy but died as a result of getting the cord caught around her neck and Marguerite attributed the disaster to the influence of a woman friend who had phoned during the birth. She remained "hostile, withdrawn and silent for many long days" (160).

> A second pregnancy put her back into a similar state of depression, anxiety and paranoia [*interprétation*]. A child was brought to term the following July (the patient was thirty years old). She devoted herself to him with passionate ardour; no one but she took care of him until he was five months old. She breastfed him until the age of fourteen months. While she was breastfeeding she became more and more paranoid [*interprétante*], hostile to everyone and quarrelsome. Everyone was a threat to her child. She made a scene over some motorists who drove too close to her child's pram. All

the neighbours objected. She wanted to take the affair to court.

When Didier was one, Élise took over childcare after he had been found sucking the grease from his pram wheel, supposedly under his mother's eye. Meanwhile Marguerite planned to leave France for America, where she would become a novelist; she would have left the child behind, but gone 'for his sake'. Instead, with both her family and her employers concerned by her state of mind, René had her sectioned for five months in a hospital at Épinay-sur-Seine.

She went to Paris in August 1925 and lived there quietly, working for the Post Office, studying English and preparing for a *baccalauréat*, which she was to fail three times. She set aside money for Didier's future and visited him frequently at first, then more rarely; but her obsession with the dangers to him continued: "'I was really frightened for my child's life', the patient wrote, 'if some disaster didn't happen to him now, it would later, and it would be my fault, I would be a criminal mother'" (163). A similar phrase appears, in Lacan's voice standing for hers, just before she goes to attack the actress: "'What would she think of me', she was saying to herself in effect, 'if I didn't come to defend my child – that I was a cowardly mother'" (172). In summer 1930, Marguerite had submitted her novel *Le Détracteur* to Flammarion, and attacked the secretary who told her it had been rejected; she was fined and told her family the cost had been incurred in a fire at her lodgings. She began writing letters to the Prince of Wales; she quarrelled with her family, grew ever more tense during her visits to Didier, taking him to and from school, "which the boy obviously did not like" (171), and telling Élise she wanted to divorce and keep the child, or else she might kill him.

What Marguerite actually said or thought, we cannot know. We certainly cannot know, and do not need to, what René and Élise thought about her; this life-story is that of Didier Anzieu, not his mother. But it is important to examine Marguerite's claims or delusions a little more, because they present a cluster of ideas that will have resonance in the rest of my argument. Whether it was realistic or justified, she wanted to achieve great things, as a scientist (specializing in chemistry) or a writer. She wrote some intense and more or less surrealistic texts, which Lacan reproduced in large sections in his thesis and refused to return to her when, as housekeeper to his father in the early 1950s, she demanded them back; they were used as the basis of a stage play, *Aimée*, first performed in 1997. She also had fantasies about a utopia that "would be the reign of women and children. They would be dressed in white. It would be the end of the reign of evil over the earth" (Marguerite Anzieu 23, cited JL 166; see also Allouch 421-36). Coexisting with this image and curiously intertwined with it is a wish to be a man, encouraged by the woman friend whom she later blamed for the death of her daughter. Lacan quotes her as saying "I feel masculine" and her friend as picking this up: "You are masculine"; this combined with a contempt for "loose women", which appears in her writings and her paranoia (JL 227). It also appears in one scene in her writing: "'I want to be accepted as a boy, I will go and see my fiancée, she will be full of thoughts, she'll have children in her eyes, I will marry her, she'd be too unhappy, no one would listen to her songs'" (Marguerite Anzieu 33). The wish to be a man is, of course, logical in the context of the ambitions of a working-class girl – see the different fates awaiting Julien Sorel and Emma Bovary (to whom Lacan compares Marguerite) or the men and the women of the surrealistic circle (see Conley). But it seems also to be what Lacan calls "a sense of psychic affinity with man" (228) which appears in an image that expresses at once the desire to know, to be and to have: "'I have so much curiosity about the male soul, I feel such attraction towards it'". We will come back to this in chapter 3 and chapter 9.

Looking back to his childhood, Didier Anzieu described himself in a draft autobiography rather dramatically as "unloved, the son of unloved people" (APP 36; see also ACD 3: "a true-false family where my father's wife was not my biological mother, my actual mother having handed me over to her sister who could not have children"). But on another occasion he characterizes himself as over-loved:

> I couldn't go out of doors without being bundled up several times over: jumper, coat, beret, muffler. The layers of my parents' care, worries and warmth never left me, even when I lived far away from home. I carried it like a weight on my

shoulders. My vitality was hidden at the core [*au cœur*] of an onion, under several skins. (APP 14-15)

There is, he says, a direct line from this to his theory of psychic envelopes. It relates, of course, to the other burden he was carrying, in the person of his dead sister – and here, in conversation with Gilbert Tarrab, it is hard to tell whom precisely he means by "my parents":

- *You are an only son, if I'm not mistaken.*
- More precisely, I'm an only child. [...] an only child is someone who receives too much, at least that was the case with me. [...]
- *What does an only son get too much of?*
- The passionate love of the parents: their ambitions, anxieties, neuroses, attention, care. I need to make clear how this 'over-investment', as psychoanalysts call it, played out in my particular case; my arrival had been preceded by that of a little sister.
- *In that case, she must have been a big sister...*
- That's true, but I meant that to me she stayed little for ever, because she died at birth. So you're right to call me an 'only son' rather than an 'only child'. In practical fact I never knew her and I lived as an only child. But in everyone's minds, this wasn't so. That lost sister, who had marked their first failure, remained for a long time in my parents' thoughts and words. I was the second one, they had to take special care of me and watch over me to protect me from the tragic fate of the elder child. I suffered the consequences of their fear of repetition. At all costs I had to survive so that my parents would be justified. But it was never enough. The least attack of indigestion or smallest draught was a threat. This put me in a difficult situation, a quite unique one. I had to replace a dead girl; yet they never let me live properly. It wasn't really a paradoxical situation; let's call it an ambiguous one. (APP 13-14)

The ambiguity, I'd like to suggest, relates not only to being preserved without being able to "live properly" – young Didier's successful development and rebellious moods soon solved that side of things – but also to the wearing of what seem to have been a *girl's* multiple skins. Let us not forget (but reserve for later) the fact that while Marguerite Pantaine was clothed in the name and burnt skin of a lost sibling of the same sex and who had lived five years, Didier her son, born on 8 July 1923, was clothed in the wrappings of a child of the other sex, who neither lived nor grew and whose skin had not yet formed beyond the womb. Neither knew the lost sibling; both were marked by this other – but differently.

While saying almost nothing about his godmother, Anzieu refers to Marguerite as – unlike him or René, both only children – born in a large family and having "tastes and intellectual gifts that I certainly inherited" (APP 16); he also describes her as succumbing, once Élise took over in the home, to "her own latent pathology" so that

> madness became a familiar reality to me early on, full of problems, heavy with menace, but a reality that I always did and still do know how to face up to firmly. Freud explains the attraction of fantasy literature in terms of its 'disturbing strangeness';[2] madness to me stood for the experience of a 'disturbing familiarity'. (APP 17)

Marguerite's on-off mothering is described not so much in terms of wrapping as in terms of feeding:

> I went from excess to lack and from lack to excess. Sometimes she'd overdo the care to an exaggerated degree, so it almost hurt, or stuff me with food that was so rich it made me throw up: no doubt she was using these physical excesses to try to make up for her inability to show affection or be loving. Other times she'd switch off, withdraw, forget I was there or needed feeding.

But either way, the maternal presence in his early life is presented as stifling, imprisoning. His father, on the other hand, also under the influence of a personal bereavement and despite an 'overflowing' love, is felt to have opened spaces up:

> If I managed to survive and grow up, despite the effects of a mother who was sometimes mad, without falling prey to the contagion of mental illness – I've made do, I think, with a normal neurosis – it's because of my solid and warm relationship with my father. He overflowed with affection. He supported me all through my childhood and later in many of my adult projects. My father lost his father at the age of twelve and until I was twelve we were close to each other. He transferred to me the best of his relationship with his own father, which had ended too early. [...] His mother died a few years later. Fully orphaned at seventeen, he [...] rose rapidly through the ranks of the Post Office up to the rank of inspector, and his desire, which I fulfilled, was to see me take as far as possible the studies he wished he'd been able to continue. (APP 18)

Father and son would take long walks or bicycle rides near Melun and talk about far-off countries, with René testing Didier on the names of the capitals. "I went round the world while walking and talking with him. I expect that explains my lack of interest in real travelling" (19).

Differently, almost contrary to Marguerite's longing for travel, frustrated by years of incarceration and the craziness of her projects, René's is internalized as mental geography, the pursuit of education and where it can take you. After some teenage conflicts and successful studies in Paris at the Lycée Henri IV and the École normale supérieure, Didier Anzieu reached the goal of psychoanalysis via these two pathways, then: the wish to "care for my mother in myself and other people" and, reading Freud, the discovery of "a new geography which took the place of the one my father had taught me. I had got a map to guide me through the internal continent" (26).

In both cases, we see an internalization of what is, originally and logically, out-there space: contained in both senses, the objects of a difficult heredity cohere in a system of thought.

During the years of education – "I got my taste for reading from my mother, but it was encouraged by my father: it was a way of bringing my disunited parents back together" (APP 23) – Anzieu sought out a series of paternal figures to guide and influence him: a young bookseller who let him browse, inspiring philosophy teachers at school who instilled in him "a longstanding vocation to teach" (24), and especially a private tutor, Zacharie Tourneur, with whom he studied and eventually coedited Pascal's *Pensées* and who disciplined his "temptation to be grandiloquent" (28). At the École normale supérieure, where he was in the post-war cohort of 1944 (admitted in 1945, after the war, a moment, according to Alain Peyrefitte, of recovering "the true, clean body" [*le corps propre*] after "four years of torture and filth – that is, so-called Aryan purity" (RK 58)), he studied philosophy but found the teaching frustrating: "philosophy had been reduced to an inferior form of rhetoric. After my mother, this was my second experience of disappointed love" (APP 29). He turned back towards his more scientific interests, in the form of psychology, taking courses in animal psychology, gestalt theory and psychophysiology. In 1947 he married Annie Péghaire, whose path had paralleled his since they met, aged 16 and 17, at secondary school, passed the *agrégation* the following year, and decided not to look for a permanent post yet, instead teaching philosophy at two *lycées* and pursuing "a more concrete education" (30) running projective tests and psychodrama groups at the Centre Psycho-pédagogique Claude Bernard, a treatment centre for disturbed children. At the same time, "I did a practical psychology course at Prof Graciansky's dermatology unit, where I gave Rorschach tests to eczema sufferers [...]; that gave me a vague early intuition of the Skin Ego".

In 1950, under the guidance of Daniel Lagache whose assistant he became in 1951,[3] he wrote his two doctoral theses, the major one on Freud's self-analysis (rather than, his first idea, on Pascal's *Pensées*) and the minor one on psychodrama (APA, published in 1956, and AA-a in 1959). He became fascinated by psychoanalysis, and had to find himself a training analyst. Having met Jacques Lacan after a lecture the latter gave at the ENS, Anzieu was flattered to be invited to be analysed by him. "My father complex again", Anzieu comments: "I let myself be taken on board" (APP 32). "I brought him a positive and intense father complex which enabled me to talk to him unstoppably and with great relief of things I had not spoken about to anybody" (RHP 245). But in another interview held ten years later, he went further, likening Lacan's failure to tell him he was the son of 'Aimée' to Lacan's failure, fifteen years earlier, to seek Marguerite's permission in relation to the thesis in which she was so central:

> No doubt he was interested in taking ENS students onto his couch, and especially the son of his former patient. He made quite sure not to tell me that 'little detail' – the fact that my mother had been the case history in his thesis – which would have meant I had to exclude him as a possible analyst. Once again, the opinion of the person concerned – my mother earlier, me this time – was by-passed. (RK 48)

Lacan later denied that he knew who Didier Anzieu was at any time during the analysis, which lasted from 1949 to 1953 (see RHP 135 and RJL 189; in both texts, Roudinesco suggests that he must have known, but may have "repressed the knowledge"). He told Anzieu the same when the latter asked him outright. For in the meantime, Annie Anzieu, "eager to meet the woman whose madness had been hushed up by her family" (RJL 189), had found his mother and she had told him about her past. Already irritated by Lacan's idiosyncratic use of the psychoanalytic framework, personal arrogance and inability to let go of his analysands, and frustrated at being unable to bring him a negative transference (see APP 33-36, RHP 245, and RK 48), Anzieu broke with him at this point, after telling him that he had written some pages of self-analysis which Lacan asked to see. "He demanded my notes: he wanted to publish them. When I promised him, falsely, that I would bring them to him as soon as I'd put them in order, I was, for the first and also the last time, in my heart of hearts [*dans mon for intérieur*], laughing at him" (APP 36).

In his heart of hearts, then, Didier Anzieu avenged Lacan's arrogated publication of Marguerite's writings by withholding his own.[4] This was not all, though. Among his complaints against Lacan is the fact that, even though the analysis had loosened the "internalized stranglehold of my parents' care and worries" (APP 35), Lacan had failed to help him analyse "the complexity of my relation to the maternal image split between an ideal and a persecutory mother", a difficulty he partially resolved in his second, more robust analysis with Georges Favez but which, he claims, he only really worked through "by a slow labour of self-analysis" (17). The mother remains inside, except where self-analysis allows access to the special kind of creativity we shall see described in the next chapter. As for the father, Didier Anzieu ended the analysis with Lacan at a time when he and Annie had just had their first child (Christine, born 1950) and were expecting their second (Pascal, born 1953). A certain emancipation from that serial need of surrogate fathers seems to have coincided with the moment of leaving Lacan, whichever we put causally first. Anzieu went on to give a courteous but spirited response to Lacan's 'Rome Discourse' in 1953, criticizing the surrealistic stress on language, and to write in 1967 an essay entitled 'Against Lacan', in which he notes: "The mother is the great gap [*la grande absente*] in the work of Lacan" (APs 182).

Among what she calls the "third generation" of French psychoanalysts, born in the 1920s, Roudinesco groups Anzieu as one of the "academics" (RHP 290), and this is where he made his career, as a lifelong educator as well as clinician. But in 1992 he himself describes a familiar ambivalence towards this world: "I have loved the academic institution like a foster-mother. Like my real mother, it was often disappointing – and doubtless often disappointed by me" (Parot & Richelle 259). Less active than many in the series of dramatic schisms that rent the French scene from the 1950s onwards, and determined never to form a set of followers unable to emancipate themselves from his influence, Anzieu nevertheless has a central place in the "hundred years of psychoanalysis" (RHP 683 *et passim*) less because of his problematic relationship with Lacan – this he shares with all the major names of his generation – than because of his unswerving independence of thought.

The Société Psychanalytique de Paris (SPP) was founded in 1926 and affiliated to the international body set up in 1910 by Freud, the International Psycho-analytic Association (IPA). After a crisis of management in 1953, Lacan, its then president, given a vote of no confidence because of his short analytic sessions, resigned from the SPP and announced the creation of a new group, the Société Française de Psychanalyse (SFP). To the surprise of the dissidents, the IPA informed the new Society that they were no longer IPA members. The following year, Anzieu succeeded Juliette Favez-Boutonier as professor of clinical psychology at the University of Strasbourg. Here he "brought with him his encyclopaedic knowledge" (Jean Muller, cited in RK 63) and developed particularly innovative courses on social psychology, group psychology and psychodrama. Active in the local community, he lectured at hospitals, prisons and businesses; in 1956 he took part in an international seminar on group dynamics run, under the Marshall Plan, by disciples of its founder Kurt Lewin (see Hubert Touzard cited in RK 70 and Chabert 1996: 17-18). His combined interest in groups and drama led to the founding, in 1962, of CEFFRAP (the Cercle d'études françaises pour la formation et la recherche active en psychologie), and in the same year he ran the first regional seminar on psycho-sociology in Murbach, Alsace. What motivated this group-centred work? To both René Kaës and Gilbert Tarrab, he describes himself as an only child "looking for siblings" (see APo 12 and APP 14). We shall see in the next chapter how this led to the theory of the 'group illusion' and the 'group ego'.

Meantime, in 1959, the SFP, anxious to be reintegrated, "began a courtship of the International Psychoanalytic Association" which was not to end until five years later (see Turkle 112). The IPA set up a committee at its Congress to decide on the French application; already in 1955, Heinz Hartmann, its then president, had written to Juliette Favez-Boutonier describing the IPA's fear that "we are seeing in Lacan the creation of a myth. And, despite your admiration for him (I know he is a brilliant man), I hope you will agree that this is more than he deserves'" (RHP 330).[5] Along with many other former analysands of Lacan's, Anzieu reported to the commission in 1961. He told Élisabeth Roudinesco:

I replied precisely and without inhibitions [*sans complexes*] to their questions about the short sessions I'd had with Lacan (without inhibitions because they clearly knew all about it). I also told them that Lacan had asked me not to mention the short sessions when I was interviewed for my application as a pupil at the SPP. The commission asked me as well to talk about one of my psychoanalytic cases. I described the most difficult one. [...] They asked me what I thought of Lacan's psychoanalytic technique. I told them that, in my experience, it had three shortcomings: an almost total lack of interpretations, an inability to bear the negative transference, and a failure to understand the specificity of the early relationship with the maternal imago. (RHP 335-36)

Despite or perhaps because of electing Lacan and Françoise Dolto to the posts of president and vice-president, the SFP was not accepted back by the IPA; instead they were given nineteen technical requirements, known as the 'Edinburgh demands' after the 22nd IPA Congress in Edinburgh at which they were formulated, as a condition for reapplying. Most ran counter to Lacan's clinical practice; article 13 explicitly demanded that Lacan and Françoise Dolto be phased out as training analysts.

Over the next two years, Lacan became increasingly isolated not only from the IPA but also from some of his hitherto loyal colleagues; in particular he felt betrayed by the younger generation whom he had nurtured – in a complicatedly maternal way, says Roudinesco (RHP 352, 518 and 620), though Anzieu would doubtless disagree – and who now turned against him, exasperated by his demands and inflexibility. In 1963, six of them signed a special motion supporting Lacan, hoping to "prevent a dogmatic split that would make Lacan into a 'charismatic leader' [the IPA's term] and the IPA into a police tribunal" (RHP 356); Anzieu did not sign because, as he put it, "his hostility to Lacan was well known and the mention of his name might counteract the moderate position they were jointly taking" (357-58). Serge Leclaire, who was representing the SFP, decided not to put the motion to the Stockholm Congress in July/August, and the IPA repeated that the Edinburgh demands must be adhered to: by

March 1964 all Lacan's current pupils must be reallocated. In October, the SFP split, Lacan was struck off from training analysis, he regrouped his loyalists around him, and the "Lacanian movement" was born (368) with the founding of the Groupe d'études de la psychanalyse, soon renamed the École freudienne de Paris (EFP).

The SFP was formally dissolved in 1964. By 1965, Lacan's group had swelled to a dominance of 134 members, the SPP had 83 and another new group, with just 26 members was founded, named the Association Psychanalytique de France (APF). Wladimir Granoff, who had fought alongside Leclaire to try to avoid schism, wrote: "'For the ex-disciples of Lacan who went to the APF, one could put it this way: "Analysis is greater than Lacan, who was my analyst but who no longer matters to me. The only thing that matters is analysis"'" (376). Among these was Anzieu. While the EFP grew with the intellectual tide that swept up structuralism in all its forms (and while it underwent further schisms, leading to the formation in 1969 of the 'Fourth group', in which women had an exceptionally strong role), the APF continued to publish and attract students, and was affiliated to the IPA in 1966 (see RHP 618-25).

Meanwhile, Anzieu's academic career prospered; he was appointed to chair the newly founded Department of Psychology at the University of Paris X (Nanterre) in 1964.

> Anzieu had a policy that he always adhered to. Faithful to the principles of Lagache, whom he admired unreservedly, on the epistemic unity of psychology, he was inspired by the wish to realize this unity at the level of both teaching and research. Curriculum development, the kind of posts he wanted to create, the staff appointed to these posts, all these depended on a long-term project of creating a balance between the various currents, methodologies and approaches in the field of psychology. Experimental psychology, clinical psychology, training psychology and psychoanalysis: all were *equally* recognized. (Jean-Claude Filloux cited in RK 72-73)

By 1967-68, the Nanterre department had over five hundred students, and Anzieu organized his staff with a stress on consultation and collegiality that was "pretty atypical at the time" (75). Anzieu wrote a short analytic account of the events of May 1968 under the pseudonym of Epistémon; it was published the same year and much read at the time and after. Dedicated to the aim of "understanding the students" (ACI 11 and APP 121-140), it takes as its premise that the academics and their pupils, for all the old entrenched hierarchies, were on the same side, "the side of non-knowledge", and that the issue was both a "generational conflict" (ACI 35) and a question of epistemology. At Nanterre, in the newly formed sections of human and social sciences, democratic methods of teaching had already been instated. What arose now was an explosive combination of infrastructural problems (too many students in a poorly equipped environment, taught by an inadequately prepared and funded staff) and the demands of a formal, hierarchical system of structuralist thought that neglected the urgent questions of history. The students were, whether they knew it or not, phenomenologists looking for a forum in which to ask questions.

Staff at Nanterre, Anzieu chief among them, were driven by the principles of the group psychologist Carl Rogers: "non-directive discussion" (33) and a democratic method of learning in which the group explores ideas in the presence of a monitor or 'animator' whose role is to guide and consolidate processes of change. By a sort of homeopathic process, the cumulative effect of these new attitudes and methods was to produce, rather than gradual change, the context for an explosion. Already at the start of the academic year 1967-8, there was a rift "between those who wanted, before going further, to consolidate the revolution inside the University by thoroughgoing structural reforms and a global, radical project of changing the very bases of society, refusing all intermediate compromises or transitory stages" (36). What emerged was something modelled more on third-world revolutionary movements, a group action which was, in the Freudian sense, 'wild'. As Anzieu explains, 'wild' refers to any psychoanalytic interpretation produced outside the regulated setting of the clinical space; its dangers are clear. In the preface to the 1982 edition of *La*

Dynamique des groupes restreints, he notes the ironic appropriateness of publishing such a study in the year in which so many French people discovered "wild" group dynamics on the streets of Paris (ADy 12).

"How", Anzieu asks elsewhere, "does a group phenomenon produce in its turn a mass phenomenon?" (APP 123). In March 1968 the students' action committee wrote to the professors of philosophy, psychology and sociology, refusing to study or sit exams; occupying the Faculty's buildings, "they also took over their symbolic role" and "launched themselves, without a psychoanalyst, into a vast collective session [...] of free association of ideas" (ACI 48-9). All students who took part in such sessions were to be awarded degrees, without examination; this dramatic emergence of the "utopian university" (49) resembling nothing so much as a "happening", a performance, a piece of "Super-New-Theatre" (70). Looking back only weeks after this declaration, Anzieu seeks to understand how and why "young people who are students represent the point in the social body where the internal contradictions of a given society become conscious" (66). Socially they are both marginal and transitory; as oedipal sons they are making the "eternal double demand for the right to power and women" (57);[6] psychologically they suffer from "an over-developed intellect combined with an under-developed emotional maturity" (113).

On 2 May, after two hundred demonstrators disrupted a French studies lecture, terrorising in particular staff who had lived through the Occupation not so long before, the Dean announced the closure of the Faculty of Nanterre. The students moved on to the Latin Quarter.

> [On 10 May], I walked around there most of the night, crossing the smaller barricades with difficulty, blocked by others. An extraordinary atmosphere of collective enthusiasm inspired these young people [...] among whom I recognized many of my own students. (73)

Anzieu's text echoes in significant ways the position of the 'group monitor' who is both inside and outside the action, observing without intervening. This position as not unsympathetic witness is represented in the text by the combination of the pronoun 'I', especially at the point when events begin to become more violent and chaotic, with a third person – "two of the four professors, Anzieu and Maisonneuve, who had arrived in October 64 and October 66 respectively" (32) – and the careful selection of a pseudonym recommended by his publisher, the name of a character from Rabelais to chime with the Rabelaisian tone of the times. The name 'Epistémon' means 'one who knows' (APP 125), but Anzieu chooses to disguise himself as someone who is "not a psychoanalyst" (ACI 54; also 111: "my own specialism: let's call it ancient epigraphy"), so that his reading of the events should be something more than a psychoanalytic one. While the students are rioting, his narrating 'I' is already working on a text:

> From the morning of Tuesday 14 March, an explosion of talk, all the talk that had been repressed before, began to spread, by chain reaction, through schools and universities, among intellectuals and artists and the liberal professions. Joint committees of staff and students started meeting in the university buildings which were occupied now without interruption by their natural users. I attended almost every day up to Pentecost weekend (1-3 June), which I spent drawing up a detailed plan for this book. After that the tension had reached such a level that I began to suffer from fatigue and my attendance dropped off. Various learned societies that I belong to had their own revolutions and I was part of that. I cut into meal-times and night-times to write, but at other times I stayed with the students as much as possible. I took part in several meetings of colleagues where we were trying to take stock collectively and think through our attitudes by confronting them directly. Those who had understood more quickly gave patient explanations to the slower ones. (75)

Two more spates of action, brutally repressed by the police, followed before the elections on 30 June, the end-date Anzieu had set for his narrative. But "the heart had gone out of it really after mid-June. [...] The students woke reluctantly out of a shattered dream" (78). And yet "the fruits of the riots were not lost". What academics need to do, according to Epistémon, is go on listening, build on the educational reforms already in place, appreciate the creative flowering manifested, for instance, in the graffiti that he quotes enthusiastically, and recognize the continuation of "a vast social experiment" (109). The causes of the failure of 1968 were both internal and external, but it enacted "the death of idols" (122) while restoring "an astonishing vitality" (123) and something resembling Erasmus's concept of enthusiastic folly.

Anzieu insists, here and elsewhere, that, though fundamentally liberal and materialist, he was never politically minded: "regimes may come and go but the unconscious remains" (APP 95; see also 54 and 92, and RK 50). But his pseudonym was quickly seen through and, far from causing him difficulties, it rather raised his profile. Edgar Faure, the newly appointed minister of education, asked him to devise a professional status for psychologists, and he worked on this during the following year; the issue was so vexed, with particular resistance from the psychologists themselves, that it was not finally resolved until twenty years later. Meanwhile, with the continuing popularity of psychoanalysis, "the number of psychology students sky-rocketed, tripling in the five years after 1968" (Turkle 171), yet their degree did not confer psychoanalytic status; confusion and frustration reigned for a generation of psychoanalytic hopefuls who still were not getting the outcome they wanted, and the Lacanian/anti-Lacanian split went on haunting the question of psychoanalytic training.

Jean-Michel Petot, Anzieu's successor in 2000 in the chair of clinical psychology at Nanterre, describes him in 1973 or 1974, during a doctoral seminar, guessing from the evidence of Rorschach and other projective tests alone not only the appropriate diagnosis but also "the family situation, the profession, the favourite leisure activity and the form of treatment (a sleep cure) undergone by a patient". "Above all", he goes on, "after amazing us by this, he took care to explain the processes of interpretation, so that divination became a transmissible technique" (cited in RK 102-3). Another colleague, René Kaës, drawing an implicit contrast with Lacan, points out that

> everyone who knows Anzieu is grateful to him for the fact that he never formed a 'school' [...]; instead of a school, he formed a 'university', assisting those who chose to enter into dialogue with him to find for themselves, within the open network of his approaches, the thread that would guide them. (RK 6)

In the early 1970s, Anzieu continued to publish books and articles, including in the newly founded *Nouvelle revue de psychoanalyse*, and in 1972 he founded two series with the publisher Dunod, 'Psychismes' and 'Inconscient et culture'. He spent two sabbatical years concentrating on research in 1973-75, and 1974 saw the publication of his first article on the skin-ego. It was in the following year that Anzieu published an article called 'La psychanalyse encore' ['Still psychoanalysis'], in which he argues that

> Psychoanalysis has become sick with its own success. [...] The problem is not to repeat what Freud discovered in relation to the crisis of the Victorian era. It is to find a psychoanalytic answer to the discontents of modern man in our present civilisation [...] We need to do work of a psychoanalytic kind wherever the unconscious emerges: standing, sitting or lying down; individually, in groups or families, during a session, on the doorstep, at the foot of a hospital bed, etc., wherever a subject is able to let his or her anxieties and phantasies speak to someone presumed capable of hearing them and giving back an account of them. (reproduced in APs 257-68; see also the extract cited by Kaës in APo 33)

In 1978, he met a painter, Charles Breuil, who, without any knowledge of Anzieu's work, had painted the image of a man wearing a woman's skin. On the back of the painting he had sketched various titles: 'Your skin', 'My skin', 'Ego-Skin', 'PO' 'Skin-ego', deciding finally on the title 'The envelope'. Delighted by this "coincidence between an artist's intuition and a psychoanalyst's idea" (APP 109), Anzieu bought the painting after the two met in 1985, and hung it next to his analytic armchair.

In 1981 both Marguerite Anzieu and Lacan died. Élisabeth Roudinesco asked Jacques-Alain Miller, Lacan's son-in-law and intellectual executor, for the return of Marguerite's writings, but he did not respond (RJL 190). Anzieu retired from Nanterre in 1983 at the age of sixty. He went on researching, with *Le Moi-peau* appearing in book form in 1985 and again in 1995; translations into Italian, Spanish, Portuguese, English and German came out in the later 1980s and early 1990s.

His creative writings, a book of short stories, a tiny volume of cartoons and a play, chimed in with continuing work on the creativity of others, from Freud to Bacon via Borges, James and, in his last years, Beckett. That book opens with a curious meeting that took place in 1989. Didier and Annie were picnicking alone in the Bois de Boulogne, sitting

> on a bench rather than the grass, too low for our won-out limbs, a few metres from the clearing where, thirty years earlier, I had read aloud from *Molloy*. On the bench we had spread out our litter, sausage-skins, melon and cheese rinds and a plastic half-bottle of wine. An old couple got out of a car; their bodies looked like the wandering shades searching along the Styx for the entrance to Hades. She had her gaze lowered, while the man's lynx-eyes darted around to see if the place was safe; you could see they were inseparable. Leaning on each other, they undertook their perambulation with wary steps. I was not sure I didn't recognize this passer-by; perhaps I knew him from his photo. Suddenly he noticed us – alarm, a flash of surprise, uncertainty and anxiety furrowed his face: *unheimlich*, a moment of uncanny recognition. Just as I was about to identify him, my wife got in ahead of me and murmured: "Samuel Beckett". I added: "and Suzanne". A moment like that carries absolute conviction. It was clear that Beckett had just seen in us a couple of old tramps straight out of his writing. For a second he had had the terrified sense of falling prey to a hallucination. [...] In the mirror that each couple held up to the other, the author thought he was seeing in reality characters he had up to now believed he had

invented, escaped from the pages of his books with terrifying powers. Finally, Beckett reassured himself of who we were: too well dressed to be real tramps. And besides, real tramps would not have recognized a writer out for a walk. (ABe 16-17).

Anzieu's delight in this uncanny moment, and the humour with which he enjoys the mirror effect of being seen as a Beckett character, leavens a book in which, elsewhere, we find the fear, not so much of death – he died in November 1999 and was suffering from Parkinson's at the time he wrote it in 1990-1991 – as of his body trapped Beckett-like in immobilization. This last book is shaped by "the project of not dying" (219): "if I stop moving, I die" (237). The main text ends at the Montparnasse cemetery, where Anzieu visits both his parents' grave and that of the Becketts. There follow seven 'postscripts', the penultimate ending "This time, it's finished. Indefinitely" (287), and the very last page giving a 'finale' which includes the lines:

> To marry the masculine and the feminine in the mind, immobility and movement in the body. To tolerate anxiety and joy, hatred and laughter. To sustain love in the gap between abandonment to the other and abandonment of the other. To foil the seductions, perversions and ruses of the death drive. To turn the negative against itself. To deny, cut, tear and transgress in order to progress. To enwrap, unfold, unfurl, unroll, curl up, interleave [*envelopper, déplier, déployer, dérouler, s'enrouler, s'emboîter*], in order to exist and coexist. To give, indefinitely, to our human finitude, a form that is never definitive. (289)

But I want to "enwrap" this chapter by returning to a description Anzieu gave, with characteristic wit, in a celebration of his seventieth birthday in 1993, of his family origin and where it placed him psychologically:

> Ten years later, Marguerite was freed. Her husband asked for a divorce. Another drama – I was going to say, psychodrama –

Marguerite's family rose up against René and scuppered his plan of legalising his union with his sister-in-law. A fresh scandal. But also, for the adolescent son, what an experience to find himself confronting the private psychosis of a mother and the neurosis of the family group! And what luck to have been sustained by the competitive three-way love of his father and two women! What an introduction to the twinned knowledge of Oedipus and Narcissus! Yes, the history of this child is the epitome of banality! (RK 50)

Chapter 2: Anzieu's theory

In 1993, Didier Anzieu considered his career, marked by its intellectual richness and variety, and noted: "Looking back on my life and work, I think I can grasp a guiding idea: unity in diversity, the convergence of parts in a whole" (RK 50). In this chapter, I want to represent the main lines of his theory, keeping in mind always that they are converging lines, whose fundamental point of connection is the image contained in his last phrase: a movement-into, a co-presence inside. What does it mean to contain or be contained? How do these processes work and what do they mean psychically? How does containment function dynamically?

Before tracing the three main areas of his theory, I'd like to begin with a concept that Anzieu proposes in *Les Enveloppes psychiques* [*Psychic Envelopes*]: that of the 'formal signifier'. These signifiers are characterized by their dynamic nature; they are concerned with "changes of form" (AEP 1); they are spatial, capturing "the psychic properties of space" (6), and represent not psychic contents but psychic containers. My interest in them is based on their potential as a way of reading cultural objects, including human figures, in terms of a linguistically perceived spatial image that embodies (this term is not casual) their particular style, manner or significance. Introduced late in Anzieu's theory almost as a taken-for-granted, they focus his own 'literary' perspective on human behaviour and motive and I borrow them as another reader of the shapes and patterns that culture thinks in.

Each individual's fundamental formal signifier derives from a time when they were not yet capable of repression, still tied in psychically to the closeness to the mother that Anzieu calls the 'common skin'. Its structure is different from that of phantasy. It takes the form of a sentence that has a subject and a verb but no object-complement; often the verb (in French) is reflexive; and the subject is a part of the body or an isolated physical form, never a whole person. It is not a scene or an enactment but "the geometrical or physical transformation of a body (in the general sense of a portion of space) which entails a deformation or destruction of form" (15); the space in which it appears is two-dimensional, and the patient senses it as external to themselves. Anzieu gives examples: "a vertical axis is reversed; a support collapses; a hole sucks in [...] a solid body is crossed; a gaseous body explodes; [...] an orifice opens and closes; [...] a limit interposes; different perspectives are juxtaposed; [...] my double leaves or controls me; [...] an object abandons me" (15-16). In the second part of this book, I will be looking, in cultural figures and artefacts, for a series of formal signifiers and how to compare and explore them. The characteristic formal signifier of Gide's desire, for instance, is twofold: 'a compulsion to empty the body of its fluid content', and 'a straight line that ends in a swerve' (see Segal 1998: 342-64), that of the public figure of Princess Diana 'a circle around, into and out of, the surface-point of the skin', that of the mode of love in *The Piano* 'the caress of the back of the hand', that of certain recent films 'desiring from inside the body of another'. What Anzieu's theory offers to such cultural readings is a structure and vocabulary that may be able, consensually, to contain them without preempting them.

Taking Anzieu's published works roughly chronologically, beginning with the two theses he first published in the 1950s, and echoing his own three-way presentation in the introduction of *Créer/Détruire*, I am going to start with Anzieu's work on psychodrama and groups; then look at his theories of self-analysis and creativity; and finally I will come to the work for which he is most famous, that of the *moi-peau* or skin-ego, and its development into a theory of thought.

Le psychodrame analytique chez l'enfant et l'adolescent [*Analytic psychodrama used with children and adolescents*] was published in 1956, and was the shorter of Anzieu's two doctoral theses. In the preface to the 1994 edition, he declares: "Like the Rorschach test as a clinical technique for individuals, the use of psychodrama in group therapy is still one of the key methods for psychologists, especially when both methods are fortified by psychoanalytic thought" (APA 1; on the Rorschach and other tests, see Anzieu & Chabert). Psychodrama is a "composite matrix of energies" (APA 7) which allows subjects "to *be* [...], to *feel* [...] and ultimately to *know* the meaning and range of what they feel" (98). Originating in the somewhat mystical theories of Jacob Moreno, it allows participants to "dramatize their individual conflicts" (99) in a group context. If it is developed psychoanalytically, Anzieu suggests that it can incorporate the key rules of the analytic setting, which are non-omission (patients are enjoined to speak without conscious censorship within a safe and permissive context) and abstinence (analysts, refusing to enter the everyday life of patients or to allow the latter to enter theirs, also refuse and forbid physical satisfactions, especially of a sexual or aggressive kind). Psychodrama "stands exactly midway between bodily expression and verbal communication" (83; see also 105). It includes a person who, in the role of analyst, monitors the action of the others, but they are part of the action too, interpreting only after the event; everyone has to take part; and touch, where it occurs in the course of the action, is monitored so that no sexual or aggressive events get out of control.

Like the psychoanalytic setting, the psychodrama is a transitional space, in Winnicott's sense of a young child's early experience of playing "alone in the presence" of the mother (Winnicott 1985: 55; see also Winnicott 1990a) which enables creative exploration of emotions and thoughts in a safe context. It is also a "container" in the sense used by Wilfred Bion, in that it gathers up the chaotic drives and emotions of infancy, as the mother does by involving the baby in her own reverie, allowing its feelings to be symbolized, voiced in language and finally thought (Bion 31-37). Again, "as in individual psychoanalysis, the balance of permissiveness and frustration produces changes in the subject" (APA 141), bringing out resistances and defence mechanisms in all participants. Everyone experiences the effects of the transference (emotions and attitudes carried over from patient to analyst) and the counter-transference (the reverse process, in which the analyst experiences emotions and attitudes towards the patient, and uses his/her experience to control and incorporate these). The key difference, and this applies to all group analysis, is that it necessarily incorporates work on "common transferential material" (APA 155). Its effect is a *"symbolic effectiveness"* (163) that both revives and repairs the participants' unconscious concerns, leading them to "emotional catharsis" (171).

Coauthored with Jacques-Yves Martin, *La Dynamique des groupes restreints* [*The Dynamics of Small Groups*] was first published in 1968. Introducing the 1982 edition, Anzieu comments:

> To human beings the small group represents a place invested simultaneously or alternately with hopes and threats. Situated between intimacy (the life of the couple or private solitude) and social life (governed by collective representations and institutions), the small group can provide an intermediary space which sometimes reinvigorates a sense of contact and sometimes helps to reconstruct the essential gaps between the individual and society. (ADy 11)

The small group, as Anzieu defines it, is different from a mob, a mass or a gang, from interest groups or associations and from large groups numbering 25-50 members, by being of a size and scope that allows each individual to interact with all the others.

In two significant metaphors Anzieu suggests that the group provides an environment which imitates in key ways the child's fantasmatic relationship with its mother. The first is the image of a mutual mirroring: "a group of equals or peers is, after the mother, the second mirror in which each one can seek an identity through reciprocal recognition" (308). In these circumstances, the group serves as a safe haven, even if it is a gang of egocentric individuals, and consensus may well reign (for other instances of such consensus, see ADy 179-81, 310 and 319). In the second, he cites the theory of his colleague René Kaës that "the space of a large group is experienced like an image of the inside of the mother's body" (41); by this he is referring to the way that members seek safety and enclosure as well as a sense of cohesion. As we shall see presently, this is true not only of large groups.

When Anzieu suggests that none of us can cope with our own aggression very well unless we have experienced the cut-and-thrust of a family group, we think of his own stormy adolescence and his status as an over-nurtured only child: "this contributed no doubt to my attraction to group psychology. One is interested in what one hasn't had. We are motivated to act and understand by whatever is missing from our life" (APP 14). More quizzically, Kaës likens Anzieu's motive as the founder of CEFFRAP to "Oedipus, an only son who wants siblings" (APo 12) - quizzically because Oedipus only finds siblings by creating them through a series of transgressions. The suggestion seems to be that the group functions, indeed, as "the inside of the mother's body" and that the events of group-space defy the 'illusion' not only of the group as a whole or its participants but perhaps above all of its creator or monitor. We can see more of this in Anzieu's next treatment of group issues, *Le Groupe et l'inconscient* [*The Group and the Unconscious*].

This monograph, which was first published in 1975, brings together all Anzieu's theories on group work. Picking up on Kaës's theory of the "group psychic apparatus" (AGI 13), he opens the 1999 edition:

> A group is an envelope that holds individuals together. [...] A living envelope, like the skin that regenerates itself around the body, like the ego which is meant to enclose the psyche, is a double-faced membrane. [...] Its inner face allows the group to establish a trans-individual psyche which I propose to call the group Self. [...] This Self is the container inside which a traffic of fantasies and identifications circulates among the participants. (AGI 1-2)

As well as looking at the structures and varieties of this psychic traffic, Anzieu pays particular attention to the role played by the monitor or monitors in a group. The analogy with the psychoanalytic cure works only partially: counter-transference is here inevitably plural and multiple. How is the monitor, as "witness-participant" (AGI 67), to take part yet remain separate enough to recognize, guide and later articulate what goes on?

In the course of the seminars, participants libidinally create a 'group illusion' in which they see themselves, idyllically centred and protected from both individual and social differences, as '"a good group [...] with a good leader or monitor"' (AGI 76). Whether they perceive the group context as a kind of body or a kind of machine, the monitor is implicitly the object of every demand. In these circumstances, an observing outsider can help the monitor to stand back from "the difficulties he has in exposing himself to the group's phantasies, whether he is too rigid and fends them off by intellectualisation, distancing himself, being 'impenetrable', or whether he is too 'porous' and finds himself invaded and dazzled [*aveuglé*] by them" (126). The skin metaphor is not arbitrary here: because the group forms itself as a psychic envelope, so its members, and especially the monitor, experience the risks of being over- or under-enclosed in their own skin. Nor is the sex/gender issue a casual one in Anzieu's account; this is why I retained the masculine pronoun in the last quotation.

Within the body that the group imagines itself to be, the monitor may play a number of roles, which Anzieu lays out in theory and in the case-studies in this book. The group may see itself on the model of the earliest "proto-group, phantasmatic, undifferentiated and reversible, the group of children inside the womb of the mother, or that of the mother in the womb of the children" (108); in this scenario, the monitor – this version is borrowed from René Kaës – may represent the good-mother pelican (a bird that is said to offer its life-blood to feed its young). A monitor who shares this phantasy may end up wanting the course never to end, in order "to keep in a state of gestation, unborn, those to whom s/he ought to give, precisely, a second birth" (112). Or the group may be conceived as what Donald Meltzer calls the "'toilet-breast'" (219, citing Meltzer 1967), into which bad stuff is thrown to be preserved. Imagining the monitor as phallic, Anzieu offers two possibilities: "the participants experience themselves as a female group possessed by the male monitor" (138) or (borrowing from Hector Scaglia)

> the power of the phallic mother is attributed to the monitor and the observer. [...] At the primitive level, which is the relation to the part-object, the relation of the newborn to the maternal breast-penis, I don't have a 'thing', but I am a 'thing', and I am it, in the best cases, for another person, as the baby is for the mother. (144-45)

In the worst cases, however, the monitor is tormented rather like Kafka's dreamlike Country Doctor: we read scenarios of humiliation and martyrdom where Anzieu represents himself as "afraid that I won't last out" (123), feeling that he is "preaching in the desert" (125) or suffering "an intense feeling of fatigue and loneliness" (128).

Part of this suffering arises out of the way he shows the group as feminine in relation to a phantasized male leader. This image of the leader goes back to Freud's two main discussions of group psychology, *Totem and Taboo* (1912-13) and *Group psychology and the analysis of the ego* (1921). In the latter of these, large-scale political groupings are united by the sense of being a host of sons each libidinally bonded to the idealized leader-figure; the latter stands as the ego ideal, a kind of paternal superego. If, for Anzieu, the analytic group feels itself to be inside the mother's body,[7] it can also manipulate the gender position of the monitor: an all-female group may "desire union with the monitor-father" and a mixed one may act as a sibling group wanting "to eat the mother-monitor in order to incorporate or become him/her" (AGI 108). In this view, the group is always feminized: "in essence feminine and maternal" (99), a sort of mouth, or "a female sex-organ, a central hole penetrated and fertilized by the words of the phallus-monitor" (188). Whether the male monitor is taken as the ego ideal or simply as the paternal imago, he must refuse the seductions of the group's demand. The group illusion is an 'ideal ego',[8] a narcissistic and feminine configuration that must be prised apart, if necessary via a controlled enactment of the "phantasy of breakage" [*le fantasme de casse*] (112-40).

In all these circumstances, the monitor is pressured by a powerful mixture of gender roles. I will return to these issues, with regard to a particular case-study, in the next chapter. A last remark is necessary here. Though Anzieu reads the group as contained inside the illusory 'skin' of its self-image, it is more precisely, as Kaës points out in his preface, a series of skins inside one other: "the group Self is a container of containers: the ultimate, irreducible kernel is that innermost individual unconscious reality" (AGI xv). And behind the paternal imago, says Anzieu, "bivalent like all *imagos* – that of the selfish, severe, cruel father whom the resentful child wants to attack but also to flee" (205), there is always "the split *imago* of the bad mother". In the non-directive activity of group analysis, both these images are in play, one behind – or within – the other. This explains, I believe, both the intensity of the group illusion and the pressures, of a particularly gendered kind, that it exerts on its monitor. So if Anzieu turned to group work in order to seek the siblings he had never had, what he found in it was something fraught with the ambivalences of his status as only son haunted by a dead sister. On the last page of this book, commenting on a case-study, he describes the situation of the subject 'Palatine', who mends her threatened skin-ego by creating a textual palimpsest to patch together the torn 'surface of inscription': "it appears that these sewn-together skins also represent the image of 'Palatine's' familial body. An only child born after a dead brother, she was a replacement piece of the family skin" (242). In a similar way, I suggest, Anzieu is patching himself a containing skin to hold together both a dangerous mother and a lost sister. We will return to this in chapter 9.

In a 1982 essay on Melanie Klein, another replacement child, and also one of that majority of psychoanalysts who are training-analysed more than once, Anzieu concludes: "Is it not true to say that the clinical and theoretical discoveries of psychoanalysts are the result of a continuing working-through [*perlaboration*] of what was left unanalysed in the course of their personal analysis?" (APs 243). We recall that at the time of the break with Lacan, Anzieu had begun a self-analysis, and he continued this not only after the happier analysis with Favez but his throughout his life; indeed, he describes his retroactive recognition of his mother's importance (remember that he became a psychoanalyst so as "to care for my mother in myself and other people") in these terms: "it was through a long work of self-analysis that I was able to reconstruct in my mind the problems of my mother's contact with me in the first months of my life" (APP 17). His last text, *Beckett* (1998), continues – and chooses not to conclude – this personal project. From one viewpoint, as we shall explore in the next chapter, self-analysis makes no more sense than parthenogenesis: the lineage of psychoanalysis insists on the production of each new analyst out of the teacher-pupil bond of at least one training analysis. From another, it is clear that, like parthenogenesis, self-analysis is a powerful fantasy: to reproduce oneself single-handed, through a mode of communication with oneself, not another. Perverse, absurd or megalomaniac, this is, for Anzieu, the base fantasy of all creativity.

The longer of Anzieu's doctoral theses, *L'Auto-analyse de Freud et la découverte de la psychanalyse* [*Freud's self-analysis and the discovery of psychoanalysis*], argues that Freud discovered – or, more accurately, created – psychoanalysis as a result of the work of self-analysis that followed a number of mid-life crises: the unplanned sixth pregnancy of his wife, a period of psychological and physical illness, a loosening of the intense tie to Fliess, and finally the death of his father on 23 October 1896. From July 1895, Freud analysed a number of his own dreams, setting this work alongside the dreams of his patients (for Anzieu always assumes at least one Winnicottian, Beckettian virtual other overseeing the act of 'being alone') and what emerged was the formation of dreams as wish fulfilments, the discovery of castration anxiety, and ultimately *The Interpretation of Dreams* (1900) and the whole theory of psychoanalysis. Anzieu in his turn analyses Freud's dream analyses, insists on the relevance of the life to the work, and derives from all this a theory of creativity.

Freud is the first example explored in the fullest presentation of Anzieu's theory of creation, *Le Corps de l'œuvre* [*The Body of the artwork*] (1981), his favourite and, he believed, least appreciated book (see APO 8). This is a study not of creativity, more often a potential than a realization, but of the act of creation, where it originates and how it is carried through into the production of a work. Familiarly, Anzieu links this to the life-cycle and to gender. His theory is derived, he announces, from three sources: himself, his patients and "contact with 'great works'" (ACO 9); and it stands in contrast to other uses of psychoanalysis in relation to creativity, which either claim to analyse "the unconscious of the text" or to give a psychoanalytic reading of the semiotics of language. His premise is that

> it is the unconscious of the author, a living and individual reality, that gives a text its life and singularity. The unconscious of the reader [...] brings to it a new life, another originality. This is the same as what occurs, in the cure, between analyst and analysand. Cut off from these two unconsciouses, the text is simply an inanimate, anonymous body, a corpus of dead letters. (ACO 12)

This is, as the first section title of the book – 'Entering into creation' (15-23) – suggests, a highly gendered representation of the psychology of the creative act. Elements of creativity belong to the five spheres of the maternal, the paternal, femininity, masculinity and indeterminate, and I shall return to these in the next chapter; but they typically take place in a male body, for

> the greater frequency of male creators is due largely to the fact that the paternal mental function is generally more developed in boys than girls, because it is the resumption, in terms of thought, of the biological function of the father, endowing the Ego with a new function, the ability to conceive codes. (ACO 83)

Thus, as aware as Anzieu is that the traditional metaphor of creativity is female reproduction – again a reason why it is traditionally ascribed to men, on the grounds that women cannot properly do both things – he sets out in this book to present an account of how male creativity is a consolidation, not a contradiction, of masculinity.

Locally, the opening moment of a creative act is a version of 'take-off' or 'lift-off' [*décollage*]. This metaphor is borrowed from Proust, in love with an amateur pilot, his secretary-chauffeur Alfred Agostinelli, who died in a accident while flying under the pseudonym of Marcel Swann; Anzieu uses it to describe the ability of the creator to 'fly above' [*survoler*] other people (ACO 17). From the longer viewpoint, the occasion is likely to be a life-cycle crisis, that of old age between sixty and eighty, middle age around forty, or youth around twenty. All these crises carry intimations of mortality, but they present themselves differently and produce different kinds of artwork: the old man seeks "'a piece' of immortality" (50), the middle-aged man a substitute for his declining potency and the solution to his mid-life depression in the form of a repaired "loved, lost and destroyed object" (53), the young man creates explosively and violently as his "work of art is an attempt to re-establish the continuity, totality, perfection and brilliance of the narcissistic envelope" (55).

All creators begin from a fantasy of heroism, generally imitative of a model, as Freud looked to Goethe and Anzieu looks to Freud, and often a model of masochistic or martyred heroism. They are encouraged in this by an adoring mother whose influence will be one of two kinds: she may overstimulate them, producing the negative effects of a second, burdensome or dangerous skin, separation anxiety, addiction or a chaotic hypertrophy of mental activity or else the positive outcome of a "totalizing-creation" (73); or on the other hand she may under-stimulate them, giving rise to a slow productivity based, like Beckett's, on meagre and desperate means, a work of "consolidation" that can last a lifetime.

The actual work of creation goes typically by five phases:

> experiencing a state of sudden shock [*saisissement*]; becoming aware of an unconscious piece of representative psychic material; raising it into a code to organize the work of art and choosing a material that can give a body to that code; composing the work in detail; producing it in the outside world. (ACO 93)

A particular creative event may not have all five phases, or it might have more; and they may be experienced at different lengths or with back-and-forth switches. The first phase is often set off by a personal crisis – loss, bereavement, illness – and is normally experienced when alone; solitude, often unbearable, being mitigated by a sort of doubling of the self, "an intense narcissistic over-investment or an enlarging of the frontiers of the self" (96) which creates a "marsupial pouch where the gestation of the work of art will occur". The ego becomes able to regress while watching itself do so; it is an extreme state, often accompanied by a sense of chill, bringing the fear of death that accompanies every birth, "a psychotic moment which is not pathological" (100); the subject feels elated, powerful and, faced with a 'break-in' through the psychic envelope, capable of "reconstituting, through the project of composing the work of art, a half-material, half-psychical skin that will repair the breach" (102).

In the second phase, a piece of repressed unconscious material appears, as in a night-time dream, but more forcefully, and the creator fixes this material "in the preconscious as the kernel of an activity of symbolization" (113). Solitude now is no longer an advantage; indeed the creator has particular need of a companion, "a privileged interlocutor" (114) – like the Winnicottian/Bionian mother who enables the child to exist in a state of 'illusion' – who can recognize, share or encourage the creative act. Thus a

[male] writer "silently addresses an ideal person, a mother or sister he would have wished to seduce, a father he would have liked to convince of his worth" (115). The third phase is the move towards finding the right code to realize the psychic material: the latter "comes in from a peripheral position to a central one" (116), and the code acquires a body. A detailed analysis of codes follows in a later chapter. The more the code and the body differ the richer the work of art: the latter may, as in the case of Flaubert, borrow from the author's own experience; or, as in Melville's *Moby Dick*, it may be an imaginary, imagoic body. The creator takes the mortal risk of pouring psychic energy out into the work – anally, they may be torn between "keeping it or making it" (122). At this point there is a conflict between the ideal ego and the superego: narcissism battles with the rule-making element that constrains it.

In an earlier essay published in the late 1970s and included in *Créer/Détruire*, 'La structure nécessairement narcissique de l'œuvre' ['The essentially narcissistic structure of the creative work'], Anzieu focuses on what he here describes as this third phase. The psychic material must be something that the individual has "left unexpressed, unappropriated, unexploited" (ACD 27), and around this "unthought, unnameable, unrepresented, unfelt thing, the work composes a skin of words or plastic or sound images" (29-30). Phallic, reproductive or anal, depending on the gender element, this narcissism is also a fantasy of omnipotence, consisting of "carving for oneself a personal act of speech [*parole*] from the common language" (38); and however megalomaniac the act, the creator fears the failure of what is, after all, also carved out of their flesh.

The fourth phase often fails to occur. When it does occur, it is of course the longest and most laborious. Willpower, determination and discipline are necessary, and what may cause it to falter is the inability to "take an internal distance from the maternal image" (ACO 126), the pleasure or symbiosis that must give way to the superego of stylistic work. In the fifth and last phase, the work, now born, must be seen off to find its own way in the world. It is difficult to end, difficult to submit to the test of others' judgment, and most difficult to confront this part-object as it proves good or bad in one's own eyes as well as those of the public. Exactly like a newborn child, "one has to present it to other people [*l'entourage*] so that it can be known and acknowledged" (ACO 130). With any work there is a "tension between the work as product and the work as process" (133), and every work contains not only elements of its creator's unconscious but also those of its potential receivers.

It is, as Anzieu himself points out, in his work on creativity that he began to develop the concept of the skin-ego (RK 34). A work of art is like a body fleshed forth, a poem is "a skin that holds together sensuality, motivity [*motricité*] and affectivity, an envelope that unifies momentarily the past, present and the expectation of a future, a membrane that harmonizes the vibrations of the body with the internal rhythm of the code" (ACO 158). Like the skin-ego, it has an inward- and an outward-looking face; but it also establishes an "empty space" (ACO 208) between the kernel and the shell of the creator's psyche, filling that space as best it might.

Anzieu's last book, *Beckett* (1998), returns to the concept of self-analysis in two ways. First it argues, as he had done earlier in a section of *Créer/Détruire*, that Beckett's writing is an extended response to the failure of his analysis with the young Bion, a soliloquy that is more exactly a free association directed to the virtual interlocutor of the invisible analyst. Half the self speaks, the other half listens, and what is spoken is "a universal message about psychic pain" (ACD 124). The book brings together, implicitly, two people suffering from Parkinson's: the mother whom it was Beckett's unconscious purpose to "*vomit*" (ACD 116) as he vomited his works, and Anzieu himself, whose relation to Beckett's literature of immobilization is, as we have seen, at once deadly and life-preserving. Intensely written and enacting repeatedly the fear of ending, this book intersperses invented dialogues between 'Beckett' and 'Bion' with anecdotes, readings and diary-entries from the three months 18 October 1990 to 15 January 1991. It is offered as "a piece of jewellery mounted on a Moebius strip that is not spatial but temporal" (ABe 13), an "immense enterprise in the service of an act of negative thought" (32). For Anzieu, who waited a lifetime to undertake it, it represents, like Beckett's *œuvre*, "a long, hard and meticulous work of composition" (114) expressing the quintessentially creative, self-analytic "project of not dying" (219).

Le Moi-peau [*The Skin-ego*] was first published in 1985 and reprinted in an expanded form in 1995. It is the best known of Anzieu's books and its influence has already been considerable, including in the English-speaking world – though Judith Butler notes that "unfortunately, [it] does not consider the implications of its account for the sexed body" (Butler 163 n43).[9] The theory is premised on the central importance of the body to psychic life. In Freud's time "what was repressed was sex" (AMP 43), in the 1980s the ignored and repressed issue was the body. Since Lacan, the stress on language had meant that the body was not being psychoanalytically theorized; yet "every psychic activity leans on [*s'étaie sur*] a biological function" (61). Anzieu's aim is to fill this gap. "Psychic space and physical space constitute each other in reciprocal metaphors", wrote Anzieu in 1990: "the Skin-ego is one of these metaphors" (AEN 58; see also AMP 28).

He begins his presentation by looking at its context. Like other sciences, psychoanalysis develops either by a "talmudic" (AMP 28) hyper-reading of a basic text or by a direct response to the questions of the age. In the late twentieth century, with a world running out of control, there is a need 'to set limits'; the typical patient is no longer a neurotic suffering from hysteria or obsessions but a borderline case (that is, on the border between neurosis and psychosis) whose problem is a lack of limits. Maths, biology, and neurophysiology have all become sciences of interfaces, membranes and borders, and embryology has shown that the ectoderm forms both the brain and the skin; thus "the centre is situated at the periphery" (31). The skin, like any shell or peel, has a double surface – "a protective one on the outside and, underneath it or in its orifices, the other, which collects information and filters exchanges" – and it is this complex structure of surfaces, rather than the old image of thought penetrating through into a truth-core, that can help us understand our physical, psychical and intellectual worlds in a different way.

The skin is "an almost inexhaustible subject of research, care and discourse" (34). The largest and heaviest organ of the body, it combines together different organs, senses, the spatial and the temporal dimension, sensitivity to heat, balance, movement; unlike the other sense-organs, it cannot refuse an impression, and we can live without other senses, but not without our skin. It appears on the embryo before the other sense systems;

its outward look varies enough to give off signs of age, sex, ethnicity and personality, it consists of no fewer than five kinds of tissue and it is the site of endless paradoxes:

> The skin is permeable and impermeable. It is superficial and profound. It is truthful and deceptive. It regenerates, yet is permanently drying out. [...] It provokes libidinal investments as often narcissistic as sexual. It is the seat of well-being and seduction. It supplies us as much with pain as pleasure.[...] In its thinness and vulnerability, it stands for our native helplessness, greater than that of any other species, but at the same time our evolutionary adaptiveness. It separates and unites the various senses. In all these dimensions that I have incompletely listed, it has the status of an intermediary, an in-between, a transitional thing. (39)

The bases for this theory lie in a number of areas: four sets of data feed into it. From ethology, Anzieu borrows the research on attachment in young mammals and humans by Harlow and Bowlby: infants cling or grasp on to their mother and when she is out of reach they show despair. In the 1960s, Harlow proved that a baby rhesus monkey preferred an artificial mother made of wire if it was cloth-covered and warm even if it did not give milk. Bowlby's research on attachment and loss and that of Winnicott, originally a paediatrician, on "good-enough mothering", stress the way in which the infant needs and uses the early relationship with the mother (or her substitute) to establish not only security but also the ability to learn to play independently (see Bowlby 1969, 1973 & 1980 and Winnicott 1990; also Montagu and Field). Anzieu sums up: "the pleasure of the contact with the mother's body and the faculty of clinging are thus at the basis of both attachment and separation" (AMP 49). Other data are derived from group psychology: in large groups people feel isolated and thus tend to cluster together or veer towards "'the skin of my neighbour'" (51; Anzieu is quoting Turquet) – from projective testing: Rorschach tests tend to produce two types of response: those that see patterns of penetration and those that see patterns of envelopment – and from dermatology: itching, blushing and other skin conditions show how well the skin may expose rather than protect.

There is also a more specifically psychoanalytic background to the concept. Freud's view of infant care focuses on feeding, but he notes the global contentment of the satisfied baby. After him, object-relations psychoanalysts use the concept of the breast to stand for the way the child introjects and projects part-objects; but Klein does not consider the whole system of infant care, and "the surface of the body is absent" (AMP 58) from her theory. Holding, in the broader sense, is introduced by Winnicott and it was left to other analysts, both British and French, to introduce the idea that the infant introjects the pattern of containment provided by the mother and such concepts as the "apparatus for thinking thoughts" (Bion), the soft or rigid ego (Tustin), the defensive "muscular skin" (Bick) the shell and kernel (Abraham and Torok) or the "mutual inclusion of the bodies of mother and child" (SAE; all cited AMP 59). Under its mother's care, an infant receives both stimulation and communication, and thus

> the establishment of the Skin-ego responds to the need for a narcissistic envelope and creates, for the psychic apparatus, the assurance of a constant, certain, basic well-being. [...] By 'Skin-ego' I am referring to a configuration used by the child's ego during its early stages of development to represent itself as an ego containing psychic contents, based on its experience of the surface of the body. (61)

Ultimately, growing out of this first configuration, "the Skin-ego is the foundation for the possibility of thinking" (62).

In its earliest days, the baby not only receives care but also gives out signals to its family circle. In the mother-child dyad it is no passive partner: by a sort of 'feedback' loop, it solicits as much as responds and can withdraw as much as be neglected: exchanging looks, smiles, noises and sense-impressions, it "acquires a power of endogenous mastery that develops from a sense of confidence into a euphoric feeling of unlimited omnipotence" (80); such over-confidence is necessary for it to move forward to further affective and sensori-motor enterprises. Babies have a "bodily pre-ego" which helps them to develop as individuals and this is based on reliable feedback from a "twinned" (81) other of a kind that will be echoed later in love relationships. As we have seen, touch is the first sense-faculty to develop embryonically and thus "the skin is the basic reference point for all the various sense data" (83); and because touch is the only reflexive sense it gives rise gradually to the reflexivity of thought. Thus "to have an ego is to fold in upon oneself" (84). The baby develops the phantasy of a "skin common to the mother and the child, an interface with the mother on one side and the child on the other" (85) – again an illusion of reciprocal inclusion that "is revived in the experience of [sexual] love, in which each, holding the other in their arms, envelops the other while being enveloped by them". But just as love can fail, the common skin can appear too tight, too loose or violently torn away, leading to pathologies of the skin-ego. We shall return to all these details.

What then are the main functions of the skin-ego, and their potential pathologies? There are eight of these, they develop aspects of the metaphor linking skin and self, and they will lead, later, to eight corresponding functions of the thought-ego.

The first is *holding* [*maintenance*]. The baby internalizes the nurturing hands of its mother, and also the way – literally in many cultures – she supports its whole body; in fantasy or reality, we all gather strength from leaning against a protective other, either their back to our front, or our back to their front. From here Anzieu sees the infant deriving the image of "an internal, maternal or parental phallus" (122) and a fear of being without 'backing' or support is the attendant risk. The second function is that of 'handling' or *containing* [*contenance*]: an envelope of sound doubles up with that of touch to provide a sense of surrounding continuity, and a dearth of such care may lead to feeling like a kernel without a shell, or to having a "head like a sieve" (125). The third function, *protection against stimuli* [*pare-excitation*; in Freud, *Reizschutz*] is like the effect of the upper layer of the epidermis; mothers provide their children with this effect and thereby save them from the need to create a 'crustacean' or muscular ego of the kind described by Tustin and Bick, or to resort to the artifice of addiction.

The fourth function is *individuation*: just as our skin represents us by its grain, colour or the signs of ageing, so we form a sense of distinctive selfhood that marks us off from others – where this fails there are the symptoms of uncanniness or schizophrenia. The fifth is *consensuality* [*consensualité*] or inter-sensoriality: "the skin-ego is a psychic surface that connects together sensations of many different kinds and allows them to stand as figures on the originary ground of the tactile envelope" (127): inter-sensoriality is a function of the ectoderm which connects the brain and the skin, so it is no accident that 'consensuality', consensus and common sense come from the same root. Sixthly, the baby's skin is treated sensually by the mother who creates it as a backcloth to sexual pleasure: *sexualization* [*soutien de l'excitation sexuelle*] is then localized at particularly sensitive erectile areas or orifices in which "the surface layer of the epidermis is thinner or direct contact with the mucous membrane produces an effect of excitation"; or the skin may be more narcissistically than sexually invested, which leads to the "brilliant narcissistic envelope" (128) of the hysteric.

The seventh function is that of *libidinal recharging* [*recharge libidinale*], maintaining inner energies and tensions in balance; its pathologies are the fear of explosion or the fear of a Nirvana-like state where all tension is reduced to zero. The eighth function is that of *inscription* [*inscription des traces*] or signification: biologically or socially the skin-ego acts as a basic parchment on which psychic incisions, tattoos, gestures, or clothing multiply like a palimpsest. And there is an extra, ninth function which, like the last fairy at the christening, crowns the positive with a negative: the skin, as conditions like asthma and eczema illustrate, also has a capacity for self-destruction or *toxicity* [*activité toxique*].

Much of the rest of the book is taken up with instances of typical pathologies, using case-studies, myths or fantasy fictions. Breathing, whimpering or vomiting are the effect of a Pandora-like containment of bad feelings that must not be let out of a box; borderline patients may find their skin-ego too stifling, or suffer a sense that their mother has torn away the common skin or forced it on them as an unwelcome, poisoned garment; without belief in the consistency of maternal care, a patient might "slip through [the analyst's] fingers" (159), unless the latter is able to reconstitute the containing function of an endangered skin-ego.

Returning to the course of normal development, Anzieu goes on to describe the 'double taboo on touching', which every child encounters and without which there is no possibility of moving from the skin-ego stage to that of thought; it precedes the oedipal taboo, which would be impossible without it. "Tactility [*le tactile*] is fundamental only on condition that, at the right moment, it is forbidden" (165). Thus the child is not allowed to touch itself or other people on certain parts of the body, is required to keep away from fires or windows and made to hold an adult's hand when out of doors. In the classic Freudian psychoanalytic setting (once Freud had abandoned the use of hypnosis, massage and 'laying on of hands described in *Studies on Hysteria*), no touching is permitted except the conventional handshake at parting; and the '*Noli me tangere*' of the resurrected Jesus represents another version of the replacement of physical proximity by spiritual distance (see Nancy 2003). The taboo on touching is double in a number of ways. It controls both sexual and aggressive impulses; it concerns both internal and external contacts and forms a distinction or interface between them, separating family-space from that of the dangerous 'outside world'; it both forbids the touch of the whole body (continued clinging to the mother) and, later, masturbatory touching with the hands; finally, it is bilateral, for its proscriptions apply to the adult who forbids as well as the child who is being disciplined. The taboo on touching is what sends us from an early echo-tactile form of communication towards a skin-ego that becomes "the space of inter-sensorial inscription" (178) and as for psychoanalysis, it uses the setting to recreate the effects of tactile contact by the very work of replacing the physical by the psychical.

Other 'psychic envelopes' are discussed in the final sections: envelopes of sound, warmth, odour and taste; and these are the key elements in Anzieu's next two books, the multiple-authored *Les Enveloppes psychiques* [*Psychic Envelopes*] (1987) and *L'Épiderme nomade et la peau psychique* [*The wandering epidermis and the psychic skin*] (1990). The final stage in the theory of the skin-ego is that of the thinking-ego.

It is not for nothing that Anzieu named his set of interviews with Gilbert Tarrab 'a skin for thought'. Just as he began his research life with the study of Pascal and his *Pensées* (1660), so he ends it with a theory that complements the psychic skin with a skin that contains – enables, controls and holds – the way we think. In psychoanalytic terms, the capacity for thought is the third term: first there is the body, then the primary process (impulses, drives, unconscious feelings) and finally there is the secondary process (consciousness, organisation, thought); it is what confronts the pleasure principle with the reality principle, deferral of gratification, acceptance of what is impossible or out of reach. We reason away our frustration, as far as we can. We mourn what is lost and transform it into knowledge. "The reality principle is a function that the ego imposes on the id" (AP 60).

The capacity to think "aspires to a unique (and utopian) universal logic" (AP 7): it is always disrupted by psychic impulses and failures. But "it is a moving moment for a psychoanalyst when a patient accedes to the possibility of thinking about him- or herself and about other people" (166).

Anzieu distinguishes between 'thoughts' [*pensées*] and the capacity for thought, the act of thinking [*penser*].

> Thoughts precede thinking. They need to be thought in order to be recognized as thoughts. They invoke the creation of an apparatus for thinking (the function creates the organ). Thinking is the part of the ego where it intersects with the mind seeking to know the object. The first object is the body; then, by analogy (in the fullest sense) with one's own body, the next is ideas.
>
> In sum, all thoughts are thoughts of the body: one's own body, other bodies; thinking seeks to bring thoughts together in a body of thoughts. (21)

Developmentally, we first think through the thoughts of others. As Bion shows, the mother 'contains' or 'digests' the child's chaotic impulses by her reverie which transforms them into "alpha elements" that can be used in dreams, symbols and conscious thinking (Bion 1984: 6-9). Both Bick and Winnicott see the work of infancy as letting the child internalize maternal care as an envelope that forms the kernel of thought – a curious play with inside outsides to which we shall return more than once. The next stage is a doubly negative one. The taboo on touching, imposed on the child, means that "putting desires and needs into action becomes dependent on putting them into words. Putting them into words becomes dependent on putting them into thought" (AP 33). But the child itself also has a capacity to negate. In conflict with its mother over feeding, the baby of about six months may spit, keep its mouth shut or move its head away from the nipple, teat or spoon; by fifteen months, it uses a consistent shake of the head or the word 'no'. Equal and opposite to the nod or smile, this gesture "marks the earliest acquisition of a system of communication" (AAc 4, citing the work of René Spitz). Once "no" is established as a word, "it is an act of *thought*" (AP 48).

The theory of thinking in Anzieu is triadic:

> the skin envelops the body; by analogy with the skin, the ego envelops the psyche; by analogy with the ego, thought [*la pensée*] envelops thoughts [*les pensées*]. Analogy is not a vague resemblance, but a term-by-term correspondence of the elements of each of these wholes. (ACP 31)

It is triadic also because there is a space between the shell and the kernel: "thinking requires exogenous stimuli (coming from other people) and endogenous stimuli (phantasies and affects) to leave in us and around us enough space to think" (AP 43). Thinking is difficult and takes energy; in order to make 'great discoveries' – here we see the connection to his theory of creativity – we need something like the transitional space and transitional object of Winnicott: a thing or a place that is both 'me' and 'not-me', an element of psychic safety that we have internalized from good early care. Triadic also, on the model of the pattern "skin, flesh, bones" (AP 7; see also AEN 63) is the structure of thinking

> the *cognitive triad*: the outside, the inside and the middle [...] In general, the triad consists, on the one hand, of two precise, opposing notions and, on the other, of the intermediate space and/or objects which are less differentiated and allow

contact, exchange, distinction and separation of the two antagonistic terms (AP 7-8).

Each of the functions of the skin-ego carries across to a function of the thinking-ego. Not to be outdone by Bion, Anzieu draws up a 'grid' setting in parallel eight functions of the skin, the ego, thinking and thoughts. They are in a slightly different order from the earlier version, and run: holding, containing, constancy (protection against stimuli), signification, correspondence (consensuality), individuation, sexualization and energization (libidinal recharging). Thus, just as the skin-ego holds the psyche up, the thinking-ego maintains the flux of thought in a 'phallic' suspension which is both energetic and consistent; just as the skin envelops the body, so thinking keeps thoughts closed or open, at the risk of losing the thread or having a "head like a sieve" (ACP 16 or AP 104 *et passim*). Just as we need to be defended against excitation, so we need not to be overwhelmed by insistent thoughts or confused by their discontinuity; and just as we register meanings on our skin-ego, so thoughts are encoded or, when this fails, become unclear. As the skin-ego achieves consensuality, so thinking becomes capable of system and coherence and we enjoy finding "correspondences between words, things and thoughts". We are individuated by our skin and our ego, and similarly we try to think with a personal "dynamic unity" – or adopt other people's systems as our own. We may think obsessively sexual or perverse thoughts. And finally, we are driven by the "epistemophilic impulse" towards "strong, authentic, paradigmatic thoughts". Each of these functions, familiarly, has its typical signifier, its principles, defences and risks – and each is illustrated by a range of case-studies.

In the 'Preamble' to *Le Penser*, Anzieu writes: "the person to whom I owe my capacity to think is essentially my father: unconditionally supporting my studies and my intellectual ambitions, he gave me two complementary experiences: the taboo and unconditional love" (AP 1).

But, as we saw in the last chapter, describing his mother in *Une Peau pour les pensées* as having "intellectual tastes and gifts that I have certainly inherited" (APP 16), he also succeeded in "bringing my disunited parents back together" (23). Referring, in 1992, to his writing in general, he concludes: "I have formed with my superego a couple united in the way a horseman is with his mount – and I don't know exactly which of us was the man and which the steed" (Parot & Richelle 257). The fantasy of unity in diversity, combined with the insistence on giving "to our human finitude a form that is never definitive" (ABe 289), seems the aptest way to close this discussion of a body of work that leads naturally to its own questioning.

Chapter 3: Anzieu and gender

Psychoanalysis, like the study of music, is highly genealogical. In *Psychoanalysis: The Impossible Profession*, Janet Malcolm quotes her source, 'Aaron Green':

> "My first analyst [...] was a very brilliant and charming old man – an Austrian Jew of the first generation of analysts after Freud – who had come to this country during the big exodus of European analysts in the thirties. [...] *His* analyst had been Sándor Ferenczi, and he idealized him. There was a bust of Ferenczi in his consultation room, together with one of Freud, who had analysed Ferenczi. I could thus trace my analytic lineage back to Freud. You smile, and you should. It's a preposterous notion. It's the most primitive kind of family romance – my parents are aristocrats, I'm descended from royalty, all that sort of stuff. I know that now. But I didn't then, and you'd be surprised by the number of people in and out of establishment psychoanalysis who hold to these childish fancies about their royal descent". (Malcolm 50)

A genealogy like this is actually conceived against the 'normal' model of dimorphic reproduction. It goes by the senior/junior pattern of education, not sexual difference. It does not take two, but one, to 'form' a pupil on the model of the teacher. However nuanced, however much training allows or engenders departure and innovation, the pedagogic chain is a fantasy of one-sex reproduction (see Segal 1998a: 1-40 and Laqueur). It flirts, sometimes dangerously, with the desire for cloning. And whereas in nature, a species having only one sex is all female, in human culture, this family romance tends to be all male: Freud, Ferenczi, Green's first analyst, Green.

Of course, as Anzieu stresses, the structure of the psychoanalytic encounter is always that of a couple, a pair, and this is one way in which, in the transference and counter-transference, variance modelled on sexual difference reappears. In this couple, as Freud put it in reference to the sexual act, we must get used to regarding the encounter as "an event between [at least] four individuals".[10] In this chapter, I want, thus, to think about the complexity of gender in three possible ways: first, the diachronic relation between generations that produces a genealogical chain on the model of either dimorphic or cloned reproduction; second, the synchronic encounter between two individuals in which each may play either or both gender roles; and finally, the gender play of the individual body, in which being whole, being 'open' or 'closed' and contending with the fluids that enter or leave the body will always play a part.

Anzieu's theory already goes some way towards a new way of thinking about gender psychoanalytically – after all, as I shall pursue later, if the skin is the 'leading organ' this makes it possible to think about bodily difference and individuation in a way that avoids the grosser gender presumptions of the penis-phallus, castration, and the rest. It is also true that a sense of human groups as containers, or of creativity as a version of parthenogenesis, carry gender implications that can be usefully developed. At the same time, attachment theory is a branch of psychoanalysis no more inclined to think of mothers as whole rational adults (I speak relatively, not absolutely, aware that no psychoanalyst thinks of human beings as whole or rational) than the other branches, and it is also the case that a field of work committed to the body metaphor might have more difficulty, not less, in emancipating itself from an anatomy-is-destiny premise. But one central aim of this book is to present a feminist view of a theory that is immensely promising for a reading of the psychic body as it is lived and read in western culture today (see Tyler 73). This does not mean underestimating the problems thrown up by a close examination of that theory but, on the contrary, taking them up and working with them. In what follows we shall find a rich mixture of angles on the questions of gender.

I shall begin this chapter by looking at a particular element of Anzieu's work, one perhaps more likely to reveal where fantasies enter or run parallel to his theory: his fictional writing. After that, I draw on gender elements of a study by Annie Anzieu, his wife and intellectual collaborator, *La Femme sans qualité* [*The Woman without Qualities*] (1989). Next, the three areas of Anzieu's theory – groups, creativity and the skin-ego – will be reexamined, each for a particular angle in connection to gender questions. And finally, I shall begin to draw out a number of issues in relation to the two questions to which later chapters will return: what do inside and outside mean for the replacement child and how can we best understand the fantasy of the 'common skin' shared with the mother?

Anzieu's short stories, especially in his main collection, *Contes à rebours* [*Countdown tales*] (1995), are quizzical and – sometimes 'darkly' – humorous. In any writing, but especially in fiction, there is, of course, no fixed point to the play of either irony or fantasy, and I am not proposing to derive an argument from the set of often sexualized thoughts contained in this collection. Anzieu offers only, by way of 'underlying philosophy', the gnomic aphorism: "Free association is, along with the horse, man's most noble conquest" (ACon 7). If we set alongside this declaration the one that closed my last chapter – "I have formed with my superego a couple united in the way a horseman is with his mount, and I don't know exactly which of us was the man and which the horse" (Parot & Richelle 257) – I think we get an equivocation about conquests and unity that might adapt quite nicely to issues of gender. The feminine is for the most part resolutely and unsurprisingly gendered 'other' in this book and elsewhere in Anzieu's work; it is the object rather than the subject of fantasies either linguistic or erotic, though at certain moments this difference is developed into another premise, one in which, in a number of playful ways, the universe is a woman.[11]

I have space here to look in detail at only three examples. The opening tale of the collection, entitled 'Inter urinam et fæces nascimur' ['We are born between urine and faeces'], opens boldly: "A library is a vagina. I mean the ideal library that gathers together the best books. Will this vagina receive me?" (ACon 11). Can "my volume" enter and be held, nurtured and cosseted in the "maternal lap of universal intelligence"? He feels the jolt of a "delicious pleasure […] that only incest can bring", wakes up in a soiled bed, aware that he has cried out, and the story concludes: "I have just been born". The clever twist of entering and exiting, being there and being made, suggests that the self is in the world as it has been, wants to be or cannot be, in a wholly feminized place: mind is body in the case of both the aspirant, dynamic self and the space, identified with intelligence as much as warmth, that is the feminine other.

'L'épiderme nomade' (reprinted in AEN) is a first-person narrative. As a child the protagonist has had the fantasy of taking off his skin at night, unscrewing his penis and leaving it on the bedside table, and of exchanging skins with another boy: "each of us slipped into the skin of the other [...] I dreamt that it was not me dreaming but him, that he dreamt my dreams in my place" (ACon 217; AEN 12-13). Heterosexuality intervenes when, at the age of seven, he spends a few days with a girl cousin three years older with whom he discovers that "the difference between male and female is not so much a question of sex as one of skin" (ACon 218; AEN 13). They take turns to tickle and pinch each other – everywhere but on the eyes and genitals – and create a bond that is reawakened six years later, during a night of soft-porn sensuality in which they lose their skin virginity while preserving the other. In a turn straight out of nineteenth-century romanticism, the girl dies a month later, having chosen this as her last act, and the narrator preserves her through an infantile fantasy: "I could not go to sleep without imagining my beloved dying in our embraces, and me laying down her body, stripping it of its skin, which I processed with the assistance of an embalmer. As long as this skin was available to my thought, she was alive" (ACon 225; AEN 19). In the morning he would imagine himself putting on her skin, and thus "I saw myself double and complete, boy and girl, her and me. So I was able to go on living".

As an adult, the protagonist displaces this fantasy into a business enterprise: he creates a macabre industry out of skin-grafts – like badges, buttonholes or keepsakes, lovers' exchanges, experiments with patches of different sexes, textures or colours – "cultivating the skin as a work of art as well as a means of primal communication between humans became my ideal" (ACon 227; AEN 21). Subscribers multiply, he makes certain rules – no children's or animal skin may be grafted, but a piece of men's and women's skin was often substituted "in the name of our basic bisexuality" (ACon 230; AEN 23). When the business finally fails, he retires, preparing for the logical death in which he will be buried in a cloak made of the sewn-up fragments of his lifetime collection, wearing "as a shroud, over my own mortal skin, this second, incorruptible skin drawn from the multitude of people I have known, who will always go with me. This cloak of suppleness, beauty and warmth will wrap me in its illusion for the long passage of eternity" (ACon 233; AEN 26).

A fantasy of being eternally wrapped in the skin of the other also marks the climax of Anzieu's last story, published in 1999. Adapted from a longer, differently concluded tale in the *Contes à rebours*, it features God woken from an endless sleep by his "clerk for external affairs" (AAI 119) who, freaked by God's terror in the face of nameless things that seem to come and go, unwisely tells him: "'Eroticize your anxiety'". This does not work: life-forms multiply with terrible speed. So the clerk advises God to transfer his excitation onto his creatures by projective identification.

> The clerk added: "Symmetrically you'll make psychotics; dissymmetrically you'll make neurotics. But first you must wrap them in membranes".
>
> God was taken aback. To help him, the clerk held out a book: *Le Moi-peau*. After creating prokaryotic organisms, God created eukaryotic organisms and made some of them male, others female.[12] Henceforth sexuality would replace God in conserving the species. The creatures became twice as excited: they were full of desires towards their similarities as well as their differences. Now the family had been created, God could sleep on.
>
> I was not allowed to wake him up to ask him his secret. But the clerk agreed to see me. He said: "God is dreaming a non-dream: he thinks he's a great naked baby feeling good in his skin".
>
> On my way out I understood why he was asleep and why, by dissymmetry, creatures dream of feeling good in the skin of someone else. This explains their failures, dissatisfaction and pain, the violence of pleasure and the pleasure of violence. (AAI 120)

Both the latter stories are based on the fantasy of "feeling good in the skin of someone else". In each case, the other people are multiple or unspecifically sexed; it is not entry but enwrapment that is desired, and the comfort of the last sentence of 'L'épiderme nomade' is supplemented by the risk and pain of the ending of 'Épilogue'. Other people's skins are both the only and the most dangerous place to want to be. We will look at this again towards the end of this chapter.

Didier Anzieu's wife Annie Anzieu published *La Femme sans qualité* in 1989; it is a combination of new and previously published essays, some highly poetic, others densely argued. Sometimes frustratingly, it develops Freud's axiom that anatomy is destiny, even though in many ways Annie Anzieu has the usual quarrels with him: his theory leaves women as failed, inadequate or negative men: "female negativity is male positivity 'hollowed out'" (AA 60; see also Laqueur 233-43, where Freud's theory is presented as the last remnant of the 'one-sex theory'). Despite this awareness, she insists on seeing both sexes through an anatomical grid: just as "it is clear that the sexual investment in the object is directed mainly in the woman towards the inside, whereas in the man it is uniquely directed towards the outside" (AA 47), so Annie Anzieu considers "a woman's psyche to be directly influenced by her sexual anatomy" (39). A "psychic cavity" (41) forms the female's almost magnetic desire for a penis to fulfil her – "the attraction of the hollow towards the object" (48) – but it is also intrinsically

> an excited space, almost entirely and immediately inside. Inside, even or especially via the detour of the clitoris. No genital that can be seen, especially by her who bears it. Only the contact of touch, feeling oneself touched and inner vibrations [...] Everything is concentrated in her internal space. (26-27)

At the worst extreme of this metaphor, a crisis for women writers is caused by the new availability of contraception and a little girl's ability to read better than a boy of the same age is the result not of verbal intelligence but of the pleasure of "playing with a symbolic phallus" (96).

But Annie Anzieu's anatomical model is also often nuanced. For one thing, inner spaces are also passages, and through passages things may go both inwards and outwards: "the double aim of coupling, pleasure and impregnation, faces the woman with the reality principle: seeking an orgasm, she obtains a child" (47). A baby may supplant the man in the woman's attention or love; but being pregnant also teaches her that every possession is sooner or later lost: "a site of exit: flows, births, waste products, things slipping out of the body. Milk, period, baby. [...] The fluid creates the conduit" (51). For another, Annie Anzieu writes poignantly of the losses of the menopause, when "the fragile surface that carries seduction falls away along with the time-lags of fecundity" (54), while the mid-life crisis of men, still able to impregnate young women, has a very different flavour. She also – exceptionally in psychoanalytic writing – represents the infant's experience of breastfeeding in a way that echoes how it feels for the mother:

> Seeking the penetration of the breast gives life, and this life is pleasurable as warmth flows again into the inside created by the feeding. Eros is set up around the edge of the lips, on the surface of the tongue, in the oesophagus with its calmed spasms. Milk sliding into the mouth or the nipple between the gums concentrates in a single point the possibility of recreating a delicious interior like the one in which the body recently bathed. (15)

Indeed for a woman incorporation has two faces and two meanings; Annie Anzieu describes this in a paradoxical passage. "Fantasies of reciprocal incorporation are inherent in femininity. Female matter is constituted in order to be incorporated by the infant when it feeds and to incorporate the male genital in the sexual act" (130). Carefully but problematically avoiding the metaphor of active and passive that haunted Freud in his uncomfortable use of 'masculine' and 'feminine' (Freud 1973: 147-50), she seems to be offering here a parallel that flattens out crucially gendering differences. Thus the act of breastfeeding – a passive penetration which allows one's fluid to be pulled into the other's body – and the woman's position in heterosexual intercourse – a process in which one is, this time, penetrated (not necessarily passively) and takes in the other's fluid yet is again deemed to be 'feeding' rather than fed – end up constituting a single kind of female (or is it feminine?) matter.

Perhaps the most important part of Annie Anzieu's argument about femininity is the last. In her account psychoanalysis is inherently feminine and analysts, whatever their sex, create an inner circulation within the space of the clinical setting that may be painful to either party, on the model of the painfulness of the bond between the mother and her foetus or neonate. This in itself is not so remarkable; what is most interesting is where she takes the observation. Reading her account of the demands of patients or the sometimes intolerable requirement to be still, experiencing "the solitude of the [analytic] armchair" (AA 109) while staying apparently calm and psychically available, one is struck by the representation it offers of femininity as a trial of strength and above all, of intelligence. In this picture of the analyst as mother the anatomical metaphor plays a more exploratory, less poetic role: it is no longer a sex-role but one of gender.

The patient and analyst are together circulating the processes of feeding, speaking and listening: "My desires in his speech. He gives nourishment to my mind. At the depths of my life the identical sensation recognizes him. Communication of drives? Association. The patient 'associates'. The analyst associates herself with the patient" (22). (I have kept the masculine gender for the patient here, as well as putting in the feminine for the analyst, since I think Annie Anzieu means the paradox of an exchange of maternal behaviour explicitly to cross sex differences.) As this image is developed, the words of both go in and out of the pores of each in the form of physical flows: "milk, blood, sperm or wind". The patient is "included with the analyst in the universe of the interior" as "analysis plunges us little by little towards the inside of words, and reaches us where words lie inside".

"The essential element of the code that determines the psychoanalytic system is the feminine element" (66). The analyst, man or woman, is the "box" into which a patient deposits her bad thoughts (37), or one of a set of Russian dolls "into which the patient projects and then takes out his dolls" (14); or again, a placenta, but one which, at the term of a period of "gestatory patience" (129), can let go the function that has shaped the mother's body as well as her child's. "The analyst as mother does not have to be perfect" (134), but, just like Winnicott's good-enough-mother, it is not enough to be devoted; s/he must also "put at the patient's disposal an adequate maternal ability to separate". Psychoanalysis is a mutual, reciprocal penetration: the patient settles into the couch "like the embryo in the womb wall" (131) but at the same time "my voice carries my words towards the space of a body" (120). And familiarly, this action may be the salvation of each – "the conjunction of my existence with that of my patient denies death" (114) – or it may destroy them. Like a doctor poking about among nasty organs, analysts do not locate the soul of their patients anywhere 'inside' unless they have let themselves be invaded:

> Analyst, what are you doing with your soul, invaded by the tortures of other people? Your mother-soul gives birth only to foetuses of life. Are you losing your own life in this persecution by identification of feelings? Or are you simply nourishing, with your generous, parsimonious placenta, someone who will leave you a little fuller of life? Or perhaps you are sustaining your own life, like a vampire, from this life poured into your ear. (141)

In this complex of mutual exploitations, a new version of Freud's 'dark continent' is proposed, capable of destruction as much as sacrifice, a potentially depressive experience, but a femininity no analyst can escape, and which requires the discipline, skills and survival of maternal intelligence (for this term, see Ruddick).

It is this necessary feminization that causes difficulties in Didier Anzieu's account of group analysis. As we saw several pages ago, he is concerned at the ambivalent role played by the male monitor in a group. In particular, his gender-role is troubled: if the group is the maternal body in which, bathed in the 'group illusion', the members want to play like little children, what happens to the monitor, another part-object who can never be simply one of this cluster of pre-sexed siblings? In the Kleinian terms that belong to this illusion, he may think of himself as the father's penis imagined inside the mother's body – but, in more classical Freudian terms, he might rather be the mother's own 'lost' phallus which, to the fetishist, can neither be admitted nor denied, and take shape as the fetish objects that stand for masculinity while appearing like femininity. Which is it to be?

Three models for groups predominate in Anzieu's account: the hive, the gang and the "proto-group" or "fantasmatic horde" (AGI 108-9). The hive is often referred to by theorists of group organization: viewed positively, it is a busy swarm of bees, negatively a termite colony that does not delight us with honey; either way it is a highly functional community dominated by females – and for this reason Anzieu, like many other scientists, is uncomfortable with it as a paradigm. "Man is not an insect, moved by instinct [...]. The problems of coordinating intelligences, of associating males together, of finding an effective balance between the inequities of individuals [...] are quite different" (AGI 48-9). The gang, paradigmatically a group of boys driven together by a sense of social exclusion, bands together as in a Barrie-esque fantasy around a "cabin" containing "treasure" (66) but, unlike the idealized male grouping contrasted to the hive, may support their unification by misogynistic acts like gang rape. Gender enters group theory in a number of guises: in the popular imagination, the mass [*la foule*] "is a woman, capricious, changeable, sentimental" (71). And groups may be structured (as in Freud) around a paternal leader-figure, the ego ideal, or around the fantasy of a maternal ideal ego, "the imago of the group's own narcissistic omnipotence (the narcissistic identification with the breast as source of pleasure and fecundity)".

The proto-group is formed by the latter fantasy, but because of this it has an intrinsically pre-sexed character: it is "fantasmatic, undifferentiated and reversible: the throng of children in the mother's womb or of the mother in the children's womb" (108). Such reversibility both goes beyond gender – pre-oedipal undifferentiation, equality, joyous fusion, comradeship – and actually presumes an essentially feminine space, the womb, in which the idyll can be held together. We remember Anzieu's formative wish to care for his mother as though she were held in his internal space. What is the place of Anzieu, or any male monitor, in the pre-differentiated womb of the group illusion? What he cannot be is a leader on the dangerous Freudian charismatic model. But he does not want to run the risk either of being devoured by their "infinite greediness" (111) or of being "torn to pieces" in the phantasy of breakage. In other words, how is he to retain any masculinity? In one case, we see this played out especially clearly.

In 1964, still working with the Lacanian model of imaginary, symbolic and real, Anzieu acted as monitor to a group of people from the psychiatric professions: thirteen in all, there were six women and seven men, the latter including Anzieu and two observers. The group's scapegoat figure was Nicolas, accused of knowing the monitor too well (he had been his student) and the atmosphere seemed, as Anzieu retells it based on an account by the main observer René Kaës, to be one of continual accusation and revolt. This account is meant to illustrate the workings of the 'group illusion', an attempt to reverse primal loss by experiencing the idealized "libidinal object" of the group (AGI 74-5). Curiously and awkwardly, the narrative is couched in a mixture of third person and first person. Thus we read: "the second question is addressed to the monitor: what comparisons has he made with other groups he has led in the past? I answer by picking up an earlier intervention..." (82). Whereas elsewhere in this book the change of narration is explained by a system of using 'I' for the self of the past and 'we' for the writing, recollecting self, here the difference between 'he' and 'I' seems less between closer and more distant modes than between a securely gendered position – even in 1964 everyone knew the monitor was a distinguished analyst and professor – and a position of uncertainty and risk in which, as it turned out, he felt humiliated by a woman.

Even thirty-five years later, Anzieu's raw anger is evident in the way he describes his rival 'Léonore'. This woman presents herself in the circuit of introductions in the first session – curiously there were, finally, thirteen sessions attended by this group of thirteen – in a way that "attracted the attention of all the men [...]: she announced that she was an expert in family planning" (77) and had already taken part in a very successful group of social workers. Léonore refuses to take orders or show respect: almost uniquely among the members, she neither knows Anzieu's theory nor intends to "acquaint herself with it now" (79). Because of her family-planning work, she is perceived by the group as "someone who has mastered the secrets of life, birth and sex". Looking back, Anzieu admits: "it seems to me now that the group gave up hope of anything from me from the moment it saw

Léonore as the source of this knowledge – the revelation of the mysteries of seduction, the primal scene and the difference between the sexes" (77-8).

These three spheres of knowledge, from which he finds himself cast out, are the fields that the group illusion tends to refuse: like unborn or infant children, the members want to believe they are without desire, exclusion or sexed identity. In horror, the monitor watches Léonore both stealing from him the position of maternal ideal ego and debasing the currency of his kind of mastery. He is "the excluded third" (78) in the couple formed by Léonore and the group.

When, in the third or fourth session, the members form into pairs, he remarks nastily: "only Léonore set her cap [*jette son dévolu*] at a partner of the opposite sex". The next day, they all decide to have a meal together, several insisting on "excluding the monitor, while of course reserving to him the chance of retribution". The monitor "interprets the dependence and ambivalence" (79) of the group towards him, and his interpretation is accepted by some and rejected by others; Léonore denies that his exclusion fulfils any desire of hers.

Dreading the end of the course, on the fourth day, the group draw a picture: an image of the Garden of Eden where, despite the innocence of the primal couple, Eve has amputated arms "'so that she can't defend herself against men's amorous attacks'". They no longer listen to the monitor at all. They announce their intention to meet later without any observers, but also insist on an extra (thirteenth) session in which they put a set of last questions – here Anzieu's first-person voice appears in each reply. In a postscript to the case-study, he takes a direct look at the relation between his own disappointment, the dominance of Léonore, and issues of equality and gender:

> What I failed to do with this group [...] was interpret their anxiety about the phantasy of the primal scene. Their refusal to talk about the question of pairings among the members, Léonore's refusal to put herself in the position of partner to the monitor, the refusal to admit that this group's existence was based on a joint initiative between the monitor and the main observer, the repeated insistence that every member was

absolutely equal, in other words, the denial of the difference between the sexes, all begin to make sense. From this point of view, the members' group illusion acted as a defence against the explanation of the origin of human beings by the sexual union of a man and a woman. The group illusion represents the unconscious assertion that groups are not born in the same way as individuals but are parthenogenetic offspring, living inside the body of a fertile, all-powerful mother. (84)

How exactly did Anzieu fail with this group, and what is he trying to do in his account of it, so fresh still a generation later? It seems to me that he was angry with Léonore for both displacing and caricaturing him, for taking away his power without admitting it, and finally for refusing, despite her egregious choice of "a partner of the opposite sex", to form a couple with him. There are thus several varieties of jealousy here – and he is frank in letting us see them in the story he tells – but the main bitterness is reserved for the fantasy of equality and in particular for the way it repudiates the social inequalities of sexual reproduction, a fantasy that is much closer to the hive. Léonore seems to have made it impossible for the group to "find an effective balance between the inequities of individuals", in other words, to "coordinate the intelligences of males associating together" (48). The gender question that this case study inescapably raises is whether man is an insect after all.

In *Le Corps de l'œuvre*, we will recall, Anzieu is clear that women have been less creative than men but reluctant to attribute this only to social disadvantages. The impulse to creativity is a 'take-off' [*décollage*] that resembles a plane or rocket lifting from earth to air. As we shall explore in Chapter 7, this is a phallic metaphor, though one that clearly plays with the risk, at the moment of erection, of another kind of taking off: castration. In *Le Penser* too, he identifies elements of thought as phallic: "trunk, tower, pillar, column, menhir, phallus: the erection of thought [*du penser*] is analogous to the erection of the body" (AP 120). At the same time, he recognizes that the making of a created object is similar to other bodily productions: "elements of the work [*travail*] of birth, expulsion, defecation, vomiting" (ACO 44; see also 56 and 81) as well as the subject-object relation of torture or lovemaking.

The gender question I want to consider in relation to Anzieu's theory of creativity is whether the ambivalence of this cluster of images is one of uncertainty or indeterminacy. By the first of these I mean a discomfiture with following the logic of body metaphors into their gender consequences, by the second a development of gender as something that fruitfully involves a body in modes attributed to the 'other' sex. As we saw in the last chapter, there is no doubt that Anzieu's creator is male: the location of creative impulses in a series of life-cycle crises is entirely modelled on a typical male pattern (though, as we have seen, his wife's work connects women's life-changes – negatively – to crises of contraception or menopause) and the son of an idealizing mother is more likely to turn out creative than other offspring (though see the later APs 229 where he suggests a creative woman can benefit from this too). At the same time, he recognizes the traditional connection between creativity and gestation. To resolve this he sets out five "forms and functions of the sexual element in the activity of thought [*pensée*]" (ACO 84): the maternal, the paternal, the feminine, the masculine and an indeterminately sexed element. I want to look briefly at each and then consider the pattern that emerges.

The *maternal* function mainly concerns the preliminary stages of creative thought: the activities of "awakening [, ...] exploration [and] classification" (84) which retrace the course of the infant's training in sensory discovery, the earliest naming of things, the rhythm of the internal organs or family life, and the reception of food and communication. All these make up a "container [of] psychic contents", and reinvoke the "training in early symbolic play, which consists of an indefinite permutation of different parts of the mother's body, both alone and together with parts of the child's body and the bodies of the outside world". All thought retains an element of the maternal in this sense, but only in order to go beyond it; thus "inspiration" (86) is a term that refers not so much to the respiratory process as to being born, coming out of a liquid environment into one of air, just as the 'take-off' is an exit into a breathable medium. The mother makes the creator male, and once he has left her this function passes to the [grammatically feminine] *œuvre*: "the creator needs the work like an indulgent mother whose committed gaze assures him of his virility".

The section on the *feminine* reiterates some elements of the maternal – "softness of contact, the suppleness of the envelope, the capacity to echo, sharpened sensoriality, the closeness of the psychic to the biological, the valuing of lived or living things" – but there are three further aspects that seem to bring new gender positions into play for the subject, or would if they were allowed self-critical development. These are:

> - the anchoring of language, or any code assimilated by the woman, in what is bodily and/or affective;
> - following from the sensation of her sex organ as a fold and then a pouch, a mental attitude that is not passive (as is too often said) but an expectation or solicitation of opening up, in the anatomical or cultural sense of an intellectual stimulation, being penetrated by a powerful idea, a project whose firmness she feels inside her, to which the woman responds actively, entering into resonance with it, taking her share, bringing her contribution, freeing her capacities, and creating from it the forms and fruits of an original sensibility and thought;
> - seeking to please others by beauty (the beauty of the body, clothes, ornaments, dance, music or poetry).

These traits, of which Anzieu explicitly says that they also "may characterize the feminine part of the psyche of a man" (87), belong respectively to the third, fourth and fifth phases of the creative process.

The other three gender-functions are related directly to the male-body idyll of Anzieu's creative theory. The *masculine* concerns "essentially muscular relations with external and sexual objects": delving, "setting up processes and tools that increase strength, skill, precision of investigation (a preoccupation with questions of method being essentially masculine)", breaking down obstacles, exploring a space by going back and forth, etc: these are at times caricaturally idealized "transpositions into the psychic sphere of the male sexual attitude". They belong to transitional moments between creative phases, in which the creator "libidinally and vigorously invests" his chosen piece of repressed unconscious material, or in which he uses the code to "set up the creative work like a machine in perfect working order which responds well to his drives". The *paternal* is the "directing thread" that coheres and enables all other elements. Just as a child only properly understands its own (or any other) origin when it inserts the paternal element into the primal scene, so the conception of people or things or ideas are all made possible only by "the introduction of a male principle into a maternal element which he makes active and causes to conceive". This principle is, like that of masculinity, perceived as phallic and magical: power, activity and particularity are its key traits. It centres and instigates creativity: "in sum, what is paternal in the work of thought is the anchoring of the message (words, sounds, images, gestures etc) in the code" (88).

The rest of this section offers two quite curious readings of the reproductive process that is, I suggest, the crucially unspoken element of Anzieu's discussion. As an instance of the paternal effect of creativity, he invokes a paradigm shift in human thought, "when Aristotle, in his *Metaphysics*, expounded that man is born of man – and not from earth, as ancient chthonian beliefs held". The image of "man born of man" - for all the conventions of French usage, Anzieu was not naïve when it came to gendered language – leads on to the last element, that of the *indeterminate*, in which "the difference of the sexes and the generations" (89) is by-passed for something that all bodies have in common, a sort of neuter grammatical form, "that pre-genital sexuality which is the anal, common to both men and women". As he describes it, the anal element of creativity is "the capacity to concentrate on a piece of repressed unconscious material, hold it tight so that it cannot escape, [...] giving it consistency and letting it ferment, push it out when necessary and, in that case, model and fashion it as it passes out"; this process is likened to the one whereby cheese is formed by "the coagulation or perhaps fermentation of milk".

This idea of the anal belongs to a pre-oedipal mode of thinking, the theory of small children who believe not only that babies are born through the anus but that they are made in the stomach. Developed here as a version of "digestive creativity", it plays a strangely infantile role in a process that has been, otherwise, consistently focused on the psychic fantasies of an adult male body. The first four gender elements in Anzieu's structure, then, all cohere in a narcissistic fantasy of the power of masculine drive, supported by and departing from a maternal base. Oddly for a theory of cultural/intellectual reproduction, and for someone who laid this as a crucial accusation at Lacan's door, "the mother is the great gap" in Anzieu's theory of creativity. Although young women are briefly set in parallel to young men as creators, even they are less likely to be successful because they lack a sufficient quotient of "the paternal function of [...] conceiving codes" (83). Again there is something perverse and curiously premodern in the insistence on conception being almost exclusively the work of a man's input - as Laqueur puts it, the traditional belief that conception "is the male having an idea in the female body" (Laqueur 1990: 35).[13] Milk is processed, things are formed, given consistency inside, and finally expelled without maternity ever appearing in the subject position, even of the unconscious fermenting process. It seems to me that there are two powerful taboos operating here. First, however curiously the metaphorical elements are turned around (thought or the code playing here a male, there a female role) the subject-position cannot remain female for more than the briefest instant. Second, the potentially male-homosexual metaphor of the creator's 'feminine' desire for the phallus, or the image of "man born of man" or the stress on the anal is never developed as part of a sexual theory of creative desire.

We see these two lacks represented grammatically in the last sentence of this discussion, ironically at the moment when gender indeterminacy is offered as its culminating point:

> The creator is he [*celui*] who possesses [...] the free play of all sexual variables in thought and who has the luck or the intelligence to have just the right variable at his disposal at the right moment [*de disposer au moment opportun de celle qui est requise*] for each phase of work. (ACO 89)

This *is* a theory of sexual indeterminacy in a positive sense - creativity relies on having the luck or wit to exploit the right gender mode at the key moment – but, as the detail has shown, it is couched in subject-object terms that cannot let go of the idea that creativity belongs only to the psychic body of a fertile heterosexual man. Thus uncertainty undermines what is almost, and potentially, a positive kind of indeterminacy.

These examples have all shown the anxieties and conflicts of a theory which recognizes the complexities of gender but finds them difficult to follow through. Through his wife, to whose work he often refers, and through his more visceral debt to a frustrated, creative mother, Anzieu must have known the ways in which female intelligence is at the heart of both the motive and the practice of psychoanalysis and in which the feminine might be understood not just as a phase or basis of male activity but something more functional. I will turn now to the ways in which his theory opens more doors than it closes. The most significant of these is, of course, the theory of the skin-ego.

All human bodies have skin, just as they all live in a world that makes them both sexed and gendered. The skin-ego offers us the possibility of a psychoanalytic theory in which sexual difference is not just, as Thomas Laqueur has it, a question of thinking of humans as being of one differentiated sex or two wholly distinct sexes, but where modes of difference are not inherently hierarchical. It is a theory centred on the body metaphor which removes castration from the position it holds in all other psychoanalytic theories. Despite the last example I gave, the penis-phallus is no longer the leading organ, and thus we are not divided by our possession or non-possession of it, however nuanced, politicized or symbolic. Thought has to do with "the relations between surfaces" (AMP 32). And similarly, sexuality depends on areas of the skin – and thus psychic skin – which are, in various ways, as the anatomical term has it, "invaginated":

> This wisely reminds us that the vagina is not an organ with a specific structure but a fold of the skin, like the lips, anus, nose or eyelids, without a hardened layer or protective cornea to protect against stimuli and where the mucous membrane is exposed and sensitivity and erogeneity are on the surface of

the skin, culminating together with a surface that is equally sensitive, that of the male gland at the tip of the erection. And everyone knows perfectly well, unless they amuse themselves by reducing love to the contact of two epidermises,[14] which does not always result in the full intended pleasure, that love has the paradoxical quality of allowing us to experience simultaneously, with the same person, the deepest psychical contact and the best epidermal contact. Thus, the three fundamental bases of human thought, the skin, the cortex and sexual coupling correspond to three configurations of the surface: the envelope, the cap [*coiffe*] and the pouch [*poche*].

Sexual relations take place where bodies meet, at the touch of two skins. All bodies have especially sensitive points. As we have seen, the sense of touch is both the earliest and the highest sense, the one that combines the rest in a consensual web, the one by which we test that the world is real. It is also the only one that is intrinsically reflexive, so that "the other sensorial reflexivities are built on the model of tactile reflexivity (hearing oneself make a sound, smelling one's own odour, looking at oneself in the mirror) and these lead to the reflexivity of thought" (AMP 84). Thus if human beings are able to feel and think, to be aware of feeling and thinking, it is not by means of a genital organ that some have and others lack - or even that everyone lacks but everyone desires - but by means of the universal human organ, our skin.

There is something potentially idyllic about this psychological premise: even though the skin is potentially the site of loss, terror and pain, it does not divide us by either gender or sexual object-choice. Containment and contact are common beginnings.

How may the skin-ego nevertheless be problematic for a gender politics? We must return to the question of the maternal. All of us start life in the womb and enter the world by taking our first breath in a space that is no longer the womb. Massaged or tormented by the passage of labour, we depart from a universe in which we are not yet capable of desire because our material needs are met without demand and in which the mother is not a person but a place. Many attachment theorists locate the initial relationship of infant to mother in prenatal life, but what changes at the instant of birth is that the place begins to become a person. Yet for many months she is a person who cannot be construed as such – for many people she is never construed as having the full span of human faculties but remains a plus- or minus-function of sufficient love. Even the good-enough-mother is actually never really good enough, because she must meet and hold the child's needs with such an ideal and unthought rhythm of provision and separation that it is enabled to do without her. The primary couple that enables a person to learn first the loop of reciprocation and then how to 'be alone in the presence of the other' requires in its senior partner, the female adult, such a perfectly subtle balance of activities and passivities that the junior, her child, will develop by coming to ignore the fact that she is there (or not there) in the same way as it ignored her when it was in her womb.

Winnicott expects the mother, in the early weeks, to be in a state of identificatory "preoccupation" that parallels the psychic needs of the neonate (Winnicott 1989: 44), and Bion describes her action of taking the "beta elements" of the child's chaotic impulses and binding them into the "alpha elements" of symbolisation, dream or thought not as intelligence but as "reverie" (Bion 36). For these reasons, perhaps, he uses the female symbol ♀ to stand for the container, an understandable choice. But he goes on to represent the contained entity as ♂. This follows the logic by which western tradition finds it almost impossible to model the mother-child couple as ♀ ♀, but insists on a ♀ ♂ pairing that it would not require in instances of authors and readers or analysts and patients (see Segal 1992: 1-11). And when it comes to it, the nipple in the mouth is another ♂ in a ♀, even though this places the mother in the illogical position of carrying the masculine symbol, on the familiar grounds that inserters are always masculine in contradistinction to feminine insertees. This is a useful instance of gender not being blindly mapped onto sex, but it follows a logic of active-passive differentiation that fails to meet the complexity of the case.

By conceiving his function as a psychoanalyst as being "to care for my mother in myself and other people", by recognizing, paradoxically, that the group illusion is that of "a throng of children in the mother's womb or of the mother in the children's womb", or by describing creativity as relying on an "internal mother" (ACO 75), Anzieu represents a different view of the containing function of the maternal. As well as offering the skin as an organ that gives no primacy to male or female, his theory develops that of Bion by taking the femininity of containment and universalizing it. Every connection between individuals, whether synchronic or diachronic – within or across generations – is a structure of Russian dolls. As Annie Anzieu observes, this makes every psychoanalytic act a structure of maternity based on both suffering and intelligence.

Before there can be a skin-ego each child must experience the delusion of a common skin with its mother: the phantasy "that the same skin belongs to the child and its mother, a skin representative of their symbiotic union [...] a cutaneous fusion with the mother" (AMP 63) in which "the surfaces of the two bodies of the child and the mother are confused" (ACD 184). This is paralleled, later, by the experience of lovers who "wrap themselves in their two imaginary maternal skins" (ACD 246) and imagine they share "a single psyche [...] a single body unique to the two of them, with the same skin" (247). In most contexts the common skin is presented as something rather difficult to visualize: an interface not unlike the Moebius ring, "she on one side, he [sic] on the other side of the same skin" (ACO 71). But sometimes it becomes something even more mysterious and similar to the paradox outlined in the last paragraph: a "phantasy of reciprocal inclusion" (AMP 85; see AGI 235). Actually the logic of the imagined fusion cannot really mean existing on two sides of the skin, for that would be the inverse of commonality. At one point in *Le Corps de l'œuvre*, another possibility is presented: referring to the mother of Borges reading to him when he was blind as she had to his blind father, Anzieu writes:

> The image of two beings in a single body. She is the eye and the hand and the breast. He is its brain and its appendix. He is the mouth that listens to her mouth. They are the same code in different languages, the tower of Babel, a total library, and all the pleasures of Babylon. This fused body and speaking mouth speak to memories, images, feelings that each of us has

> preserved more or less obscurely out of what was a decisive stage in our individual psychic origins.
>
> When I ask myself why I get such pleasure from reading and rereading Borges, I think of a poem in *El Hacedor*, which is rarely cited [...] and whose ending sums up for me what he seems to me to have been trying to say throughout his writing, his crucial experiences as a child and story-teller, the operation by which the code, any code, allows one to order and communicate everything one has felt of one's own body and that other body from which one's own is not yet quite differentiated, the body of the mother, the body of a garden, the body of a suburb. (ACO 316)

The "body from which one's own is not yet quite differentiated" suggests another version of the common skin – here again, "a library is a vagina" - the prelapsarian space of life in the womb, where one skin indeed did service for two people. Rather than an interface, this is the kind of containment that Anzieu sees as fundamental to creativity: thus Borges "lets us share in the joyful wonderment of the child who discovers that his body coincides with a code and that he can play with that code just as his mother and he have each been playing with the body of the other". If we remember that his theory of creation began and ended with the self-analyses of Freud and Beckett, we have another instance of the rerouting of a metaphor: self-analysis is not parthenogenesis, it is something more like the playing of the transitional space in which the child thinks itself alone and is not – but not even that, it is the mutual play of reciprocal inclusion, both in the same skin. What does it mean to think one is alone? Probably something rather like the 'formal signifier', where we sense an image that is bodily yet untied to a body, of ourself yet external to ourself, essentially spatial but unlocated in space. This formal signifier is tied to the idea of the common skin but has no overt knowledge of it; minimal in every sense, but all the more powerful, it is a psychic form distinct from the idea of containment but presupposing it.

Creativity and fantasy - existing inside the skin of someone else. With this we return to the image of Anzieu's first and last fictions. Where this comes from, I think we can guess: not so much a fantasy as the reality of the replacement child.

Marguerite Anzieu began life as the replacement child for a sibling of the same sex who had died of burns, and whose skin was thus unimaginable as a place to be. Even her name was the same, and she grew up surrounded by family members who had known the other child and seen her die. She was, indeed, the last daughter of her family, all the children after her being sons. As we saw in chapter 1, she felt both identification and attraction towards "the male soul" (JL 228) which suggests an unsettled relationship to gender that, however normal for an ambitious woman, might also be connected to the doubled burden of the same-sex replacement child. Didier Anzieu was born as the replacement of a child of the opposite sex, unnamed, whose death by cord strangulation was unpredictable after a successful pregnancy, the culmination of his mother's fears and the start of her paranoid terrors about his own safety. The gender significance of replacing your own sex and the other sex will be explored further in chapter 9; for now, I want only to observe how variously we can carry another's lost body around, upon and inside us. These other bodies are gendered and must gender us.

We have seen, then, how the genealogical chain carries one demand, the meeting of pairs another and the illusion of being a separate body a third. In the second part of this book, we shall see many more examples of how touch and gender are embodied as formal signifiers of desire.

Chapter 4: Gide's skin

In this second part of the book, I will be looking at instances of cultural figures and objects in which the sense of touch, the relation to skin and various kinds of 'formal signifier' can help us think about desire and experience. The first two chapters in this section each take an individual figure, one man and one woman. In rather different ways, they exemplify how desire functions both in self-directed ways and in a circuit involving same-sex others. Both could rightly be called narcissists, seeking reflection in others in determined, public and acute ways, but also isolated in their particular versions of visibility. One is homosexual and the other heterosexual, but the circuit of desire is curiously counter-intuitive: in Gide, whose sexual practice was intense if not obsessive, there is, finally, a sort of renunciation of the male other, a 'swerve' of undesire at the end of every chase, leading to lonely stasis; in Diana, whose failed fairytale is paradigmatic of a peculiarly contemporary breakdown of the heterosexual compact, we find an exchange of sympathy that is essentially woman-to-woman. Femininity is at stake - and at fault? - in both their stories.

I begin with four quotations by or about Gide which, in seemingly very different ways, represent his relation to skin. The first is about itching:

> The itches I've been suffering from for months now (or rather, for years, with interruptions) have recently become intolerable and made it impossible for me to get any sleep at all in the last few nights.
>
> [...] *On both legs, from the ankle to the calf, and on my arms (but here less violent and more spread out), it's like a incessant provocation to scratch myself until the skin is worn off. One resists only with the greatest difficulty and by perpetual efforts; I force myself not to give in, knowing by experience that the relief is minimal and having, last time I let myself do it, given myself a nasty wound.*
>
> Actually, nothing shows on the outside: it's just underneath the skin, like a poison trying to escape - an injection of extract of bedbug.

> [...] it seems to me that a real pain would occupy one's attention less and altogether be more tolerable. And a real pain is, in the scale of misfortunes, something more elevated, more dignified; itching is a petty form of suffering, impossible to own up to, ridiculous. People have sympathy with a person in pain; someone who wants to scratch himself just makes them laugh. (GJ2 263-4; the section in italics was omitted from the *Journal* published in Gide's lifetime)

We shall see in a moment what Gide did with this difficult symptom.

The second quotation is about clothes. It comes from the memoir written from 1918 to 1951 by Gide's friend, apartment neighbour and grandmother of his daughter, Maria Van Rysselberghe, known as the *petite dame* [little lady], and it describes an indoor outfit Gide wore in March 1924:

> a brown velvet jacket, camel-hair waistcoat and trousers. I can see him now, stretched out on the large yellow sofa in the drawing-room, cosily embedded among the cushions; no one could be more supple, more nobly nonchalant; but he is incapable of casualness, of just being carelessly casual; he cannot let himself sink into comfort, he settles himself in it carefully, cautiously, finding the very best position for his limbs and his clothes, always taking his pleasure with deliberation. (CPD1 194)

Next, from Gide's autobiography or coming-out book, *Si le Grain ne meurt...* [*If it die...*] (1926), two paragraphs in which he describes his mode of desire. In the first paragraph quoted here (which appears thirty pages later in the text than the second one), he is referring to his horror at an act of sex between a friend of his and an Arab boy:

> As for me, who can only understand pleasure when it is taken face to face, reciprocally and without violence, and who often, like Whitman, get satisfaction from the most fleeting

> [*furtif*] contact, I was horrified both by Daniel's act and by seeing Mohammed accept it so obligingly. (GJ3 312)
>
> [...]
>
> there are some people who fall in love with what resembles them; others with what differs from them. I am among the latter: what is strange appeals powerfully to me, just as I am repelled by what is familiar. Let us say also, more precisely, that I am attracted by what remains of sun on brown skins [*par ce qui reste de soleil sur les peaux brunes*]. (GJ3 283)

My final, longest quotation is about pleasure. In the Journal of Gide's friend Roger Martin du Gard, the latter reports a conversation of May 1921 in which he is told about Gide's unusual sexual 'disposition':

> Gide needs to empty himself out completely of sperm, and he reaches this state only after coming five, six or even eight times in succession. I don't need to mention that there was no trace of bragging in his account: far from considering this phenomenon as an 'exploit', he seems to see it as one of the 'freakish' aspects of his nature. First he comes twice, more or less at the same time, without a stop, "like a singer", he said, "who takes a second breath in the same song to sing another note, with scarcely a perceptible pause". "The second orgasm", he went on, "seems to climb on the shoulders of the first..." These two orgasms, the first two, which are quite clearly differentiated, characterized by two distinct ejaculations, one immediately after the other, apparently astound his partner. But that is not the end. The third one follows soon after. He can rarely come more than three times with the same person. When circumstances permit, he then finds himself a second partner and comes the fourth and often fifth time. After that he is in a very special state, in which it is impossible for him to feel any further desire for those two partners, but at the same time he urgently needs to ejaculate more in order to reach a point of satiation, with all the sperm

emptied out. And normally he can only reach that final stage after he's gone home, by masturbation. It is quite exceptional (he says this has happened no more than ten times in his life) that he finds the partner attractive enough to exhaust his need for ejaculation with him. So when he leaves the place where he's come three, four or even five times in less than an hour, he goes away unsatisfied. He is in such an unbearably tense, uncertain, unsatisfied state that he can think of only one thing: getting home and masturbating as many times as it takes to reach the point of total emptiness. However much he reasons with himself, tells himself it's putting his health at risk, he considers that this emptying out, however harmful it might be, is nevertheless better, infinitely less prejudicial, than the state of unsatisfaction he would otherwise be left in.

I asked him a few specific questions. I wanted to know if this repeated orgasm wasn't simply a single one divided up into several successive moments. But no. Each time he came, the quantity of seminal fluid was like a normal orgasm. The quantity might be a little less the last few times, but the total amount of seminal fluid produced is the same as a normal man would shed if he came once each day.

Gide told me that he had wondered about this problem all his life and never managed to explain it. He had found only one possible explanation: the fact that as a child he had tried repeatedly to cure himself of masturbation, especially at a period of his life when this self-discipline was mixed up with a religious prohibition. During that period, which went on beyond his adolescence, the idea of *sin* was fixed so powerfully in his mind that he felt he was committing a lesser wrong if he did not reach a complete orgasm. Confusing the sin with the actual ejaculation, he had acquired a real skill in onanism, so that he could reach the exact point where he had almost reached orgasm but the ejaculation, the spasm had just not yet started. This series of shockwaves without culmination, which had become a regular habit with him, may have been the cause of the peculiarities of his disposition today, for he would repeat them often, sometimes a whole night long, without ever allowing himself to commit the 'full sin'. (Martin du Gard 1993: 232-33)

These four observations span the years 1921-31 (Gide was born in 1869 so this is more or less his fifties) but they describe him from adolescence to the beginnings of old age. To sum them up: he experiences his own skin as something that irritates; he needs to clothe and 'live in' it with sensual care; he desires the skin of others as an object of difference, glowing with the darkness of seemingly absorbed sunshine, and yet, despite this desire at the level of the skin's surface, he is satisfied by the most fleeting - or did he really mean 'furtive'? - touch; he also has a relation to pleasure that is never satisfied, that can only briefly find an end to unpleasure when he has completely emptied the bodily shell of its fluid content.

Let us go back now to each of these aspects in turn and see where and how they form a single picture.

Itching is, as Gide sharply observes, a mode of suffering that is difficult for others to take seriously. If pain is tragic, itching is comic. In Gide's fiction, Catholicism is viewed largely with a comic, Protestantism with a tragic eye. This is why itching reappears in his only fully comic text, *Les Caves du Vatican* [*The Vatican Cellars*] (1914), in the person of the amateur knight errant and virgin husband, Amédée Fleurissoire, who is plagued, during his journey to save the supposedly kidnapped Pope, by boils, bugs and other figures of the pathetic grotesque written on the skin. Anzieu comments on itching:

> Pruritis is not only linked to guilty sexual desires in a circulatory play between auto-erotism and self-punishment. It is also and primarily a way of drawing attention to oneself, more especially to one's skin insofar as it has not received, in early childhood, contacts from the mothering environment that were gentle, warm, firm, reassuring, and above all,

> significant [...]. The itch is the itch to be loved, above all by the beloved object. [...] Generalized eczema may represent a regression to the infantile state of complete dependence. (AMP 55)

To Anzieu, who 'psychologizes' skin conditions in a way many might find reductive, itching is a plea. Other authors look instead at itching and related skin problems – eczema, psoriasis or leprosy (see Connor 2004: 227-56; Detambel 91-5; Jobling 1988, 1992 and 2000; on leprosy, see Richards and William Ian Miller 154-57) – in terms of their creation of stigma, a response that is internalized by psoriatics and thus, extending their body boundaries, circulates to include their families and carers, who experience "stigma fallout" (Jobling 2000: 100). The shedding of living tissue seems to cast 'dirt' on the environment in an uncanny way, spreading the effect of an over-visible body that is also felt as tainted from within. Shame and horror are particularly acute with skin disorders. Even if the actually itching is caused, as Jobling asserts, as often by the treatment as the disease (1992: 344), the fact that it is a symptom that seems to make the sufferer laughable rather than tragic is part of the problem.

In Lorette Nobécourt's *La Démangeaison* [*Itching*] (1994), itching changes from agony to pleasure in the course of a life lived in open repudiation of the family, institutions and society, as the protagonist Irène assumes (in the existential sense) the isolation that her surface appearance makes compulsive. In an interlude without symptoms, she experiences "the real but astonishing joy of feeling identical to other people" (Nobécourt 78), but when the symptoms return, she resumes – and sexualizes – the alternative joy of rejecting "the social comedy. The rest of them were healthy more by terror than by choice. I held on to my illness like a definition [...] I knew I was lucid, bound for a state of pure interiority, close to indifference, towards an intolerable toxicity" (94).

The toxicity of the skin-ego is Anzieu's ninth, supplementary function, its ability to destroy itself. All skin diseases present on the visible surface of the individual the kind of poisoning that we identify, internally, with cancer, that we once connected with the cannibalistic, 'consumptive' effect of TB, and that we more recently associate in a mode of unstoppable flow with the postmodern nightmare of AIDS (see Sontag). The phantom of leprosy – to which I will return in the next chapter – is the most traditional of such stigmatizing skin conditions. In Leviticus, what has been translated as leprosy (Hebrew *tzara'at*) was almost certainly psoriasis (Jobling 2000: 99). Leprosy causes more acute disfigurement of the face and other extremities, affecting the voice as well as skin-colour, and itching may be no part of it, but what it shares with other conditions is the separation of the affected individual from other people – for centuries, by a complex body of laws based not so much on fear of contagion as on "religious implications" (Richards 53).

For Gide, it is possible that his itching is part of the horror – he called it by the Goethean term of *Schaudern* [shuddering] – that seemed, on a few childhood occasions, to open the flood-gates of an unspecific grief, and made him cry: "I'm not the same as other people! I'm not the same as other people!" (GJ3 166). Being different shows on the body. Everyone is different, of course, but some people seem to etch out difference voluntarily or involuntarily in a mutual attack between person and skin.

What Gide does with his symptom of itching is turn it to comic effect in fiction. But the comedy ends in pathetic – not tragic, but more disturbing – mode when Amédée comes into contact with the desirable skin of a beautiful other. Amédée ends as the martyr to Lafcadio's *acte gratuit*: if the latter sees a light in the landscape before he has counted up to ten, he will push this silly old man out of the train window. A light appears, and out Fleurissoire goes. Here, by contrast, is Lafcadio's skin, from an early sketch for the character:

> On his arms and thighs and all down his back from the nape of his neck to that point (the axis) where Greek statuary places the bouquet of curls of the satyr, he had kept a silky golden-blond down which his mother, laughing, called his milk hair, as the first teeth are called the milk teeth. She loved to see him naked and, far from being shocked, was amused by his immodesty. (GR 1573)

The beautiful, wild child – or what Leslie Fiedler calls in *Love and Death in the American Novel* the 'good bad boy' – has a skin that, far from breaking out in embarrassing hives or spots, seems to carry forever the glow of the egg. Surprisingly perhaps in Gide, in whom desire normally focuses an adult male gaze onto the body of a boy, and in which the only scene that represents adult-child desire as dangerous and corrupting is that of an aunt trying to seduce a nephew (in *La Porte étroite* [*Strait is the Gate*, 1909] – we shall return to this), here it is the mother whose admiration sensualizes the surface of the son. What we can see here is the sort of solution that Anzieu envisages: a pure skin is made possible by a good maternal gaze. But in Gide the gaze is sexualized: it depends on a kind of maternity that elsewhere, and painfully, he will associate with seduction.

Gide's own mother had, by the account of his lifelong friend Jean Schlumberger, a face covered in marks and moles, one of which scratched when she tried to kiss you; his own warts and sensitive skin seem to have been, as Anzieu would agree, doubly created by her (Pierre-Quint 432-33). As Gide presents her in *Si le Grain ne meurt...*, she combined an excess of maternal solicitude, a love that, on the death of his father, became a second, inescapable skin – "I suddenly felt myself enfolded by that love, which from then on closed in on me" (GJ3 138) – with the stern, unyielding attitude of a censoring superego. But shortly before he died, he wrote a homage to her, describing her as something much more nuanced: shy, reserved and quietly disappointed at her husband's failure to appreciate her attempts to be attractive. In this version of her body, she almost exactly incarnates the inversion theory of Karl-Heinrich Ulrichs, the *anima muliebris virili corpore inclusa* [a woman's soul enclosed in a man's body], with the ironic addition that, unlike Proust's Charlus or other willing or unwilling male displayers of the signs of homographesis (see Edelman), for a *woman* who hides her femininity under a 'mannish' shell, there is no possibility of coming out.

It was Gide's mother who also encouraged his love of fine material next to the skin – in both a positive and a negative way. Negatively, he blames her for the purgatory of a youth spent wearing starched shirts which formed an excruciating breastplate invisible under his jacket:

> My mother took great care that no expense she went to for me should let me know that our fortune was considerably larger than that of the Jardinier family. My clothes, identical in every respect to those of Julien, came like his from *La Belle Jardinière*. I was extremely sensitive to clothing and it was a real torment for me to go about dressed hideously. What heavenly bliss it would have been to wear a sailor's outfit with a beret or a velvet suit! But neither the 'sailor look' nor velvet were to Mme Jardinier's taste. So I wore skimpy little jackets, short trousers, tight at the knee, and striped socks which were always too short and either crumpled down around my ankles like wilting tulips or disappeared into my shoes. I have saved the most ghastly thing for last: the starched [*empesée*] shirt. Not until I was almost a grown man was I able to persuade my mother not to have my shirt-fronts starched. It was the done thing, the fashion: there was no way out. If I finally got my way, it was purely because the fashion changed. Imagine the unhappy child who, unknown to all the world, winter and summer, at school or at play, wears hidden under his jacket a sort of white breastplate ending in an iron collar; for (for no extra charge, no doubt) the laundress would also starch the neck-band to which the collar was fastened. If – as was the case nine times out of ten – the collar did not fit exactly, it formed excruciating creases, and if you happened to sweat, the shirt-front was quite unbearable. Just try doing sport in that get-up! The whole thing was completed by a ridiculous little bowler hat... Oh, how poorly the children of today realize their good fortune! (GJ3 133-34)

Positively – as one might think – this treatment inspired him when grown up to pay particular attention to the quality of fabric in the flowing capes and scarves, soft hats and djellabahs that his wealth (now freed from mother's moral proscriptions) enabled him to choose. It is this that creates the "noble nonchalance" Maria Van Rysselberghe was to admire. But it also led to an appearance that other people often judged eccentric or peculiar, to cite Roger Martin du Gard's description of their first meeting in 1913:

> his eyes are hidden under the brim of a battered hat; a vast cloak hangs from his shoulders; he looks like a scrawny out-of-work old actor [...] or maybe one of those old hacks from the Bibliothèque nationale, professional copyists with none-too-clean underwear, who doze over their manuscripts after lunching off a croissant. (Martin du Gard 1951: 12)

His son-in-law Jean Lambert was later to quote Gide's tailor as telling him "quite emotionally about his taste for fine fabrics. He had three or four indoor outfits: a black velvet suit, a jacket of beige wool with wine-coloured lapels, a white flannel jacket that was put on him the day he died" (Lambert 51), but also to poke gentle fun at his obsession with the right underwear: "every one of his friends has some vest or underpants story to tell about him" (149).

Indeed it was the interface between outer-wear and skin that most preoccupied Gide. Roger Martin du Gard threatened to walk out of a cinema when Gide announced to him in a stage whisper that he had underestimated the heating and now urgently needed to remove one of the two pairs of pants he had put on (Martin du Gard 1951: 117). Pierre Herbart, whose first meeting with Gide was marked by the view of an ugly wart on the hand of which, afterwards, Gide denied all knowledge, describes this and similar moments of obsession:

> Gide is the slave of his body's every whim. If he's cold, he asks the friend who is driving him to dinner in the country to make a detour of 100 kilometres to pick up the sweater he has left behind and which is the only one that will do. If he's hot, he asks Roger Martin du Gard who has gone to the cinema with him to help him take off his underpants. His strange ways of being comfortable (he sleeps fully dressed, has to have cotton blankets and pillowcases without borders) seem so uncomfortable that people think him spartan. But he has to

have them at all costs, even if it is at the price of everyone else's comfort. He is quite unaware of the petty tyranny he imposes [*qu'il fait peser*] on those around him and seems genuinely surprised when they reproach him. He has so few needs! (Herbart 47-48)

'*Peser*' is precisely the word: this way of imposing his peculiar and specific requirements on his friends is surely a belated revenge on those undershirts whose starched surface tormented him, like the essence of bedbug, in a way that could not be perceived by other people. Soft and shapeless fabrics give him a good exterior; underneath everything is constantly to be remade, rearranged, accounted for.

It is in *Les Caves du Vatican* again that the clearest example of the moral-material relation of body to clothing (first and second skins, imposed and desired skins) is to be found. Way back at school, under the influence of the endlessly disguised – thus endlessly visible *and* invisible – Protos, Lafcadio acquired a system of belief that divides the world into those whose skins are labile and flexible, 'the subtle ones' [*les subtils*], and 'the crustaceans' [*les crustacés*] whose rigid carapace marks them out as patently unimaginative, clumsy, unmovable. It *is* a question of internal versus external skeletal structures, in other words, of what can be seen and what cannot; and there are two cardinal rules: "1. the subtle ones recognize each other; 2. the crustaceans do not recognize the subtle ones" (GR 855). Beauty, however, overrides this kind of visibility. It is not long before the unrecognized Protos (temporarily at least) takes his revenge on the peachy Lafcadio, whose beauty is such that it cannot be ignored or disguised.

Of the beauty of 'Émile X', another golden boy, whom he met at a Paris bath-house, Gide comments:

> He swims extraordinarily well, and nothing, I think, gives the muscles such rhythm and harmony, strengthens and lengthens them, as swimming. Naked, he looks magnificently natural; it is clothed that he seems uncomfortable. In his working clothes I would hardly recognize him. Doubtless it is also his habit of nudity which gives his flesh that smooth matt gleam. His skin

is blond and downy all over, and at the dimples of the sacrum, in exactly the spot where classical statuary puts the faun's bouquet of curls, the light down gets a little darker (GJ1 328)

Swimming creates the solid body unclothed except in liquid. We return here to the significance of fluids for the Gidean body. As a child, Gide's imagination feasts on the fantasy of becoming Gribouille, a storybook character who falls into a river and turns deliciously into a water-plant; his lifelong fascination with the sensuality of fluid on flesh, in bath, sea or shower, is noted by his doctor and first psycho-biographer, Jean Delay (Delay 520-39; see also CPD2 106-07).

The sun-on-skin effect can be found, then, not only in the 'other world' of North Africa, but also in the hidden places of Paris where nakedness and the rough clothes of the tailor's son exchange places. This Journal entry dates from 1902, the year of the publication of *L'Immoraliste* [*The Immoralist*], a fiction in which the hero emerges from the shadowy corners of an academic upbringing and an almost fatal case of TB via an intensive narcissism of the skin which makes him become visible to himself and, by the identical move, invisible to his concerned (and thus infantilizing) wife. This is a novel full of good blood (that of the bright-eyed Arab boys) and bad blood (the feminizing, debilitating blood of TB that Michel bizarrely passes to Marceline via a miscarriage). But there is a twist behind the tale of sunny-skinned boys and wronged wives: Gide's cousin-wife Madeleine had the same dark complexion as her Creole mother, and – as we shall see later – fluids play an essential part in Gide's troubled representation of the two versions of femininity that these two women signify.

In the fourth of my quotations we saw a peculiarly compulsive fantasy: Gide's need to empty the body of its fluid content in the form of repeated orgasms, until it has been released from the 'itch' of desire by complete emptiness. Anzieu describes an obsession with emptying-out [*vidage*] in his discussion of Pascal in *Le Corps de l'œuvre*. But Pascal is the exact inverse of Gide, because his is a fear of his body flowing away, combined with a terrified "aversion for water" (ACO 323). Anzieu relates his phobia to the sight of his pregnant mother's weighted body. Pascal was a delicate child: "all his sickly life he would suffer from an emptiness in his body. All his scientific life, he would search, in the figures and elements he studied, for a centre of gravity" (ACO 324). Gide, also physically weedy, and fascinated by science but dedicated to writing, reverses the equation: he seeks to create emptiness, a vacuum, inside himself.

What does it mean to become skin and bone by ejecting everything liquid out of the body-self system? In Gide it is part of a more general concern with casting-off or ridding the self of its possessions (see Segal 1998a: 47-50); but this version is specifically a fantasy of the body. I want to look first at the endpoint of the process, its teleology; then I will turn to a more general theory of bodily fluids and the hydraulics of desire.

Gide discovered the term 'anorexia' late in his life and understood it in its broadest sense of a failure to desire or "absence of appetite" (GJ3 994). As we shall shortly see, however, he was anorexic in today's sense in his early teens. This was just one of a series of neurotic symptoms which functioned as the 'plea' Anzieu identifies in eczema and other skin conditions. This inability to eat is another version, I suggest, of the compulsion to empty himself which is anorexic rather than bulimic – we shall see in relation to Princess Diana how bulimia is one of a series of feminizing circuits – because it is a trajectory, not a circuit, a trajectory that moves insistently in the direction of the ideally emptied body.

Maud Ellmann likens anorexia to the political protest of self-starvation, both, however improbable it sounds, containing elements of *"jouissance"* (Ellmann 13). Fasting may have an aspect of aggression or competition that is, very like potlatch, directed towards others, or it may (as on Yom Kippur) be the joint act of a community. But it is also erotically self-directed. In the way Gide describes his sexual practice, with a sort of non-arrogant astonishment, it is an entirely inward-turned activity. One of the most striking aspects of it is the serial irrelevance of the companion: the first, desired boy and the second, less attractive boy play the role almost of witnesses or pretexts rather than objects of desire; it is clear when they are no longer required, and then the endgame is a *corps-à-corps* between self and self, hand and flesh. This is, then, an instance of the way that, in Gide, desire swerves away from the other, and turns to a quite different purpose.

In Kafka's writings, there is an idealization of skin-and-bone (for his attitude to his own, extremely thin body, see Gilman 1995: 65 and

Anderson). Near-tragic protagonists are finally shovelled out with the rubbish – our anachronistic drive to connect his imagination to the reality of the concentration camps is understandable but misdirected. Quintessentially, Kafka's hunger artist is *proud* of his ability to starve.[15] The trajectory of this story is a movement away from public display to private enactment and, along the way, from the frustrations of visible success to the sanctity of unseen failure. It begins with a public performance, followed with excitement and admiration by the town; at this stage the hunger artist is presented in a plural or generalized third person. Bit by bit the general aspect of the story focuses down onto one eccentric individual for whom forty days are never enough and the euphoric ceremony of release from the cage is unbearable because he longs to stay there:

> why did they want to rob him of the glory of fasting on, not only in order to be the greatest hunger-artist of all time – he was probably that already – but to beat his own record, surpass himself to a point beyond comprehension, for to him there was no limit to his ability to starve. (Kafka 396)

The limit is set, finally, not by the spatial end-point of his skin but by the temporal deadline of collapse, and then only because the public no longer bothers to stop, look and tot up his achievement. At this point he is found under a heap of straw and beckons over the inspector to whisper (as best he can) in his ear: he was not really an artist after all, but only made an exhibition of himself with this fasting "because I couldn't find any food that I liked" (403).

This impossibility – or inability – is of course at the heart of Romanticism: very sensitive people cannot consume the world they find themselves in because they are, in evolutionary terms, without a fit ecology. As in so many of Kafka's other texts, the last page opposes the rude beauty of a healthy creature to the pathetic death of the martyr-hero. For all its genuine ambivalence, the ending does not abandon the metaphor of making art out of physical self-depletion.

Can we see Gide's sexual practice in a similar light? The very possibility of repeating the pattern desire-repeated-to-depletion every couple of days speaks of the inexhaustible replenishment of bodily fluids. But the desire to be skin-and-bone suggests something else. Pleasure is unpleasure because it aims at the expulsion of itself.

Much has been written in recent years about the significance of body fluids. Anzieu notes the importance of the relation of the psychical skin to a fluid economy: "thinking in economic terms (accumulation, displacement and the discharge of tension) presupposes a skin-ego" (AMP 60). Freud's economic theory of desire – what I shall henceforth call the hydraulic theory – relies on the concept of fluid circulation in a closed system of constancy (whether derived from Fechner or Claude Bernard, see F-Ego 387 and AP 60), according to which the ego serves as a "reservoir" or "storage-tank" for the libido (F-Ego 407), a "displaceable and indifferent energy [...] employed in the service of the pleasure principle to obviate blockages and to facilitate discharge" (385). The skin with its orifices forms the outer edge of a circulatory system of tubes and channels seeking both stasis and outlet. It is surely clear that this models desire on the supposed functions of a man's body. Drives drive fluid down channels towards a point of outlet; desire pumps out fluid; then he is at rest. The ideal post-orgasmic state, the end-point of masculine – that is, in Freud, all – desire, is a blissful stasis of undesire in which all the fluid levels are steady and release has provided the answer (the scratch) to which tension was the question (the itch). In hydraulic heaven, no vessel holds more than another.

Any hydraulic system has three main defining characteristics: first, its closed and containing structure of canals; second, connected to this, the finite quantity of what is contained and directed; and finally its teleology of balance and stasis. Nirvana is reached at minimal tension, when there is perfect rest. When movement begins again, it depends on a choice of this way or that, a binary of either/or (sexual release or repression; neurosis or sublimation) and also on a concept of flow which is finally gravitational. When Édouard in *Les Faux-monnayeurs* [*The Counterfeiters*] advises Bernard: "it is good to follow your inclination, so long as it goes upwards" (GR 1215), he is uttering a hydraulic paradox.

The fascination of body fluids is that they do not obey the hydraulic theory based on scarcity, but on the contrary, they replenish themselves by a logic of plenty. They are fascinating for another reason too. Where they breach the bounds of the body - shed or shared, sweated or consumed, through wounds or orifices - and prove the skin-ego's grievous inability to contain, they become both dangerous and ambiguous (see the classic text Douglas especially 121; also Kristeva, Héritier-Augé, Laqueur, Grosz, Shildrick, and William Ian Miller). Like dirt - defined as "matter out of place" (Douglas 35) - these fluids carry all the contagions of pollution; but they are feared especially because they represent the ambiguity of gender. Just as the distinctions of gender are both fixed (at any time) and arbitrary (socially and logically), so the significance, value and even the gender of a body fluid will depend on whether it is found inside or outside a body marked male or female.

As recently as the seventeenth century, Europeans believed that women's blood and men's semen were of the same, transformable substance. Thomas Laqueur describes the interconvertibility of body fluids in the one-sex model of the human body. Among the ancients, there was "a single general economy of fluids driven by heat" (Laqueur 41), in which "like reproductive organs, reproductive fluids turn out to be versions of each other" (38); in Renaissance medicine too, "semen was part of a more general traffic in fungible fluids" (101). What this shows is that differences which may be proposed as absolute in one context are understood as relative and exchangeable in another. Without this assumption, there could be no ideas of pollution or contamination; with it, the danger can be seen to reside always in the coefficient of femininity - the negative gender - that may be absorbed, caught or passed from one body to another. One of the ironies of Gide's desire to be emptied of sperm is how it exemplifies his horror of the feminine.

In many ways the empty body which is the goal of Gide's desire is the perfect instance of Freud's hydraulic theory. Here the very evocation of excess is understood in terms of a principle of 'enough'. Gide wants to be emptied, purged and cleared of that internal irritation of desire which makes him uncomfortable. Evacuation, to find peace. A compulsive need to finish with what produces un-peace. I want to end this chapter by looking at two moments in Gide's writing where the fluids that concern him are explicitly feminine ones, belonging to those he saw as the worst and best of women, but travelling, for good or ill, very near to the male body that encounters them.

These three characters appear side by side in two places, the fiction *La Porte étroite* (1909) and the autobiographical *Si le Grain ne meurt...* (1926). In both, the story is told of an adolescent boy returning uninvited to his uncle and aunt's house one evening: his uncle is out, and he sees, through the open door of his aunt's brightly-lit bedroom, a scene suggestive of her adulterous desire; then he goes up another floor and finds his cousin, her eldest daughter, kneeling, weeping and praying by her bed and, as he understands her suffering and vows to protect and comfort her all her life, his own life changes direction irrevocably.

I have given this story in the barest of outlines, because the two versions are built up, otherwise, of rather different elements. In *La Porte étroite*, written soon after *L'Immoraliste*, during the first few years of Gide's marriage and an intensive period of secret pederastic activity, the aunt is presented as a sulphurous, fascinating figure, always lolling on sofas, reading poetry, dressed in scarlet and glimpsed, during this 'discovery scene', as smoking and laughing in the company of a silly young lieutenant. Most importantly, the scene immediately before this one shows her calling her nephew Jérôme over to her, commenting on the boring way his mother dresses him, and then

> [she] slips her hand inside my shirt-front, asks with a laugh if I am ticklish, moves further down... I gave such a violent start that my shirt tore; my face ablaze, and as she called after me: "Pooh! the silly fool!", I fled, running down to the far end of the garden, and there, in the little tank of the kitchen garden, soaked my handkerchief, applied it to my forehead, rubbed my cheeks, my neck, everything that woman had touched. (GR 500)

Whether this scene is historical, as Lacan and other psycho-biographers have supposed (see Lacan 1966, Jadin 1995 and Millot), or a representation of the dangers of paedophilia distanced by Gide via a change of sex, as I believe (see Segal 1998a: 282-98, and Segal 2000: 346-62), it gives a particular cast of meaning to the encounter between boy and girl that follows it. Lucile Bucolin's disturbing seductiveness is directed specifically at her nephew via his skin: she tickles and touches, he runs away and tries to wash her off (for a telling parallel, see Derais and Rambaud, discussed in Segal 1998a: 331-41). While these scenes are not contiguous in time, they follow immediately on the page. After sensing an aberrant sexuality through the door of her room (her other children are playing at her feet: this again suggests the corruption of the young), he moves to Alissa's closed, dark bedroom, and there, as he stands beside her kneeling figure, "I pressed my lips on her forehead and through them my soul flowed forth" (GR 504).

What exactly has happened here? First it was as though a contagious fluid had passed, via the boy's fascination and the woman's seduction, from aunt to nephew: Jérôme felt polluted and sought a way to cleanse himself. Nothing has of course literally flowed from one body into the other: as in most pollution theories, contact is enough for the object to seem sullied, and even the fresh rainwater rubbed all over his skin cannot clear her off him because of the familiar psychology of paedophilic sex whereby the child's sense of complicity, however ill-based, always adds shame to anger. In the scene that follows, the boy's decision to protect his cousin "against fear, against evil, against life" is conditional on her ability to cleanse him of her mother's bad substance. What this must mean is that she takes that substance on. If it flows out of him onto her skin while he is filled with the milk of human succour, what is it ever going to be possible for Alissa to be? Readers of the novel will know how she starves to death by a process of cultural and moral attrition that is carefully left obscure.

In the other version of this 'discovery scene', published in his memoir nearly twenty years later, after a crisis in Gide's marriage and a waning of his supposedly protective attitude towards Madeleine, the fluid dynamics of tears on skin have changed subtly. There is no preceding scene of attempted seduction of boy by woman; the aunt is still presented as a vaguely transgressive figure, but without any specifics and with no suggestion of paedophilic interest. All he sees through the door of her room is her lying down being fanned by her two younger daughters. The uncle is (again) out of the house, but so apparently are the sons, and there is no lieutenant. Instead, the house is full of women.

In the absence of any seduction scene, what appears immediately before this encounter is a description of young André's own anorexia and his cousin's playful reaction to it – it is, incidentally, striking that this is the only place anywhere in his writing where Madeleine is represented as a mischievous child.

> I ate little, I slept badly. [...] I had no taste for anything, I went to the table as one walks to the gallows; I only swallowed a few mouthfuls by dint of great efforts; my mother implored, scolded and threatened; almost every meal ended in tears.
> [...] I lived alongside my cousin in a conscious community of tastes and thoughts which I tried with all my heart to make ever more close and perfect. This amused her, I think; for instance, whenever we were eating together at the rue de Crosne, she would tease me by preventing me from having my favourite dessert by denying it to herself; she knew quite well that I would not touch any dish unless she had some first. (GJ3 158-59)

Here there is no sultry aunt and nothing André has been dirtied with. What we have been shown of his infant 'wickedness' (on the opening page of this *Ur*-coming-out book) is an innocent masturbation spontaneously generated and neither learned nor shared. This small vice has had large effects: his mother weeps, his father dies. He does not do that any more but he is still a cause of his good mother's tears when she tries in vain, with all the other women, to make him eat. Anorexia gives him perfect power over her, but not over Emmanuèle (his preferred name, here and in his Journal, for Madeleine). Rather, this boy – his thoughts and tastes – is filled up by this girl. Let us see what then happens when he finds her kneeling beside her bed in the darkened room. "She had not got up. I did not realize at first that she was unhappy. It was only when I felt her tears on my cheeks that all of a sudden my eyes were opened" (GJ3 160).

This time, then, it is not his soul but hers that pours forth, and the fluid passes out of her onto his skin, wept down his cheeks. Remember: this is the moment of *his* discovery of female sexuality, not of hers. Having already helped him to starve himself, she takes what in this book is the exclusively feminine fluid of her mother's sexuality and weeps it onto his face. What he gains is a vocation; what is left inside her is the cruel duty to become a woman who will have henceforth "to judge her mother and condemn her conduct" (GJ3 161).

Gide wrote after Madeleine's death that she was never meant to be Alissa – or, to be precise, that Alissa was never meant to represent her. After all, Alissa never marries Jérôme and Madeleine finally agreed, following Gide's mother's deathbed wishes, to marry him. But there is no doubt that in later life, many years into their unconsummated marriage, she became very like her. She did not starve herself literally but withdrew into Cuverville, the family home in Normandy, while Gide lived mainly in Paris, or travelled, surrounded by friends and the second family, the Van Rysselberghes, among whom he chose to have a child.

We have seen that it is only when Emmanuèle's/Madeleine's tears run down André's cheeks that his anorexia loses its compulsiveness. At the moment that he infers her horror at women, her refusal of her own mother, he becomes able to mother her – this is the only scene in which she is described as a child. She enables him by that (seeming) agreement – we shall speculate in a moment about how far it can be understood as hers – to become her container. Containment conceived as protection is the conscious part of the story: he will provide the shell to shield the "intolerable distress" that she has hitherto carried alone. But she also gives him a destiny. Of course he cannot let her go after that, and of course her resistance is a sort of logical outrage that cannot be accepted. On the less conscious level, there is the risk that the soul she constitutes inside him might still have too much negative body. Her distress after all has the shape of the bad woman. So Madeleine must become the virgin wife – a wife being no-body – so that she shall never become either her own or anyone else's mother. The consideration that her body might have had its own desire is possible only after she is dead.

Only if Madeleine stays still (sexually and geographically) can he move – not simply because the relation of self to umbilicus must be made constant if he is to go so far without ever going too far, but also because he can continue to contain her only if she is not full. She must not fill herself up because her repletion would be his diminishment. She must be emptied also because in this way she can provide a different kind of container for him – not a skin but a frame. In *La Porte étroite*, we can see what is created out of the gradual depletion of the woman. As visitors to Cuverville can verify, the gate at the bottom of the garden is not narrow at all: on the contrary, it is almost square, like a frame.

In this image of the misloved woman as an empty frame contained inside the man who alone can 'know' her, we may recognize both the theory from Winnicott and Bick wherein the developing child internalizes maternal care as an envelope that forms the kernel of thought and also Anzieu's own conviction that, as an analyst, he cares for the "threatened and threatening" mother within him. What is contained, in each case, is a vulnerable and fearsome container, held in suspense (contained) within what is effectively a 'common skin'.

The elements that make up the girl Madeleine find their definition, then, in the moment when she is 'recognized' as full of nothing but suffering, the motherless child by choice that the boy has not yet found a way to be. Let us return to two fundamental images in Gide – first, the man emptying out his internally-accumulated fluid in the presence, serially, of two male others and himself; second, the young woman weeping her mother's contagion onto the face of the young man where it anoints him and empties her. In both cases, fluid feminizes and is feminized as it flows out (see Miller 1997: 103); then each emptied figure becomes its true unencumbered sex – the man cleared of all his desire, spare and strengthened; the woman drained of hers, pure and "resigned" (see GJ1 573; also CPD1 11). Both hunger-artists, both nothing but skin.

I'd like finally to go back to the first quotation, the scene of itching. Remember, now, the wicked aunt and her tickling fingers? If we look again at the terms of the text – particularly the bit Gide chose not to include in the Journal published in his lifetime, italicized below – we can see the similarity of terms between this and another instance of temptation, resistance and futility:

> it's like a perpetual temptation to scratch myself until the skin is worn off. One resists only with the greatest difficulty and by perpetual efforts; I force myself not to give in, knowing by experience that the relief is minimal and having, last time I let myself do it, given myself a nasty wound.

> the idea of *sin* was fixed so powerfully in his mind that he felt he was committing a lesser wrong if he did not reach a complete orgasm. Confusing the sin with the actual ejaculation, he had acquired a real skill in onanism, so that he could reach the exact point where he had almost reached orgasm but the ejaculation, the spasm had just not yet started. This series of shockwaves without culmination, which had become a regular habit with him, may have been the cause of the peculiarities of his disposition today, for he would repeat them often, sometimes a whole night long, without ever allowing himself to commit the 'full sin'.

Let us suppose, then, that the internalized superego which forbade masturbation, the mother who, embodying both parents, pronounces and represents the 'double taboo on touching', were conceived to have made pleasure both impossible and unstoppable, like an itch that you can never scratch enough, then, if you follow the "incessant provocation" of that temptation of the skin, what you might create in yourself is the horror of a nasty wound, a breach in the skin-ego through which something like feminine desire might just begin to pour out or in.

Chapter 5: Diana's radiance

In her interview with Martin Bashir for BBC's *Panorama* programme on 20 November 1995, Princess Diana describes bulimia thus: "You fill your stomach up, four or five times a day – some do it more – and it gives you a feeling of comfort, it's like having a pair of arms around you". What this image suggests is the curious image of the body feeling embraced *from the inside*. She adds immediately "but it's temporary" – and indeed it must be. This point of balance is achieved momentarily and regularly with bulimia; as we have seen, the rhythm of anorexia is a gradual, inch-by-inch continuum towards an ideal end-point of emptiness, each stage being a slightly increased ratio of refusing self to matter barred entry, but bulimia – and this is what makes it an easily kept secret, a structure of visibility that can remain strictly invisible – is not a continuum but a circuit. The circle it describes has four points, two states and two actions: the excessive ingestion of bingeing leads to fullness (negatively 'the bloated stomach', positively 'the embracing arms'), this in turn leads to the excessive expulsion of vomiting and its temporary resting-point is an emptiness which, unlike the emptiness of anorexia, is a version of stasis equal and opposite to the moment of comfort in that it brings a pleasure close to orgasm. To borrow Diana's own words again: "I remember the first time I made myself sick. I was so thrilled because I thought this was the release of tension" (Morton 56).

If the moment of being emptied is pleasure, the point of greatest fullness is one where the skin 'fits' exactly so that the sense of being held is a snug relation of content to surface, balanced on that fine line between satisfaction and disgust. Unlike the Gidean or Kafkaian ideal self of skin-and-bone, it is a filled self, reminiscent of Anzieu's description of how asthmatics do not want to let their breath go: "asthma is an attempt to feel from the inside the envelope that constitutes the corporeal ego: asthmatics inflate themselves to the point where they feel from underneath the frontiers of their body, to assure themselves of the enlarged limits of the self" (AMP 130). Here, for the bulimic, the limit point is reached, the internal meniscus breaks and guilt brings it all up again. Referring to a time, much later, when she had got over her bulimia and become able to believe the adulation of a grateful crowd, Diana comments: "I can now digest that sort of thing whereas I used to throw it back" (Morton 67).

The bulimic circuit flirts with the anorexic continuum by trying out at every fourth moment a relationship of surface and depth in which emptiness makes the wrapping ideal – that is, more surface than depth, more container than content. But turn the circle two more points and fullness proves that the container can also be not so much ideal as perfect – what Anzieu calls "the sac that contains and retains inside it all that is good and full" (AMP 61). The bulimic's skin demonstrates its dual function by holding in and keeping out by turns, inviting sensation and protecting against it. This is the basis of the circuit of bulimia – a circle around, into and out of, the surface-point of the skin.

In *Les Enveloppes psychiques*, Anzieu compares anorexia and bulimia: "anorexic disgust and the irrepressible greed of bulimia" (AEP 132). But disgust is as much part of bulimia as greed, and it is not greed, in any normal sense, that motivates the binge-eating, but the drive to circulate food without possessing it. Rather than consumption, this seems to be a fascination with repeatedly rehearsing consumption without being its slave. The slavery of bulimia, conversely to the very different slavery of anorexia, is reproductive of itself; for this reason, if for no other, it is feminine. The bulimic of either sex is repeating the pattern that relegates women to reproductive rather than productive work; but it is not work in that it has no end-product; the body disguises its self-disgust in a 'normalizing' treadmill of giving and taking.[16]

Radiance, surprisingly perhaps, works in a very similar way. How frequently and fascinatedly commentators use the term, along with a plethora of cognate images, to describe the 'Diana effect' we shall see in a moment. Radiance ought to mean a system of light (or heat) emerging from the interior of the person; thus we speak of the secret radiance of pregnant women or the eyes shining as the 'windows of the soul'. But more exactly we are thinking of a circular system in which what comes out has first been put in. Only our gaze makes her look radiant. She is, as Martin Amis put it, "a mirror, not a lamp" (Amis 53). Rilke describes this exactly in the second Duino Elegy when he likens angels to mirrors that "draw their own streamed-forth beauty back into their own countenance" (lines 16-17). With radiance, like bulimia, the circuit presupposes a breachable surface that lets the inner out and takes the outer in, but it also depends on an ideal state of the closed, gleaming face: "'She's so beautiful', breathed the crowds in Australia during her first tour. That skin! Those eyes! Ever after, people who met her in person remarked on her radiance. It was as if, they said, she was lit from within" (Tennant: 'Incomparable queen': 3); or again, in the words of a doctor in a Moscow hospital, "she was a wonderful person, it was written on her face" (Fox et al 10).

"An instant radiance" is said to have emanated from Princess Diana "when she walked into the room". Put another way, "she glittered, and the glittering sucked you in". She had a "stellar luminescence", was "shining [and] golden", "illuminated our lives", was both "gentle [and] flamboyant", "flashed those sapphire eyes" and "radiated warmth". Commentators from all political sides have compared her to a light source: a "beacon" (Margaret Thatcher), a "comet" (Simon Jenkins), a "shining light" or a "bright star" (Lords Archer and Hurd), a "crescent moon" (Simon Hoggart), a "paper lamp" (Maya Angelou) and the "sunshine" (Donatella Versace and a correspondent in the *Bury Free Press*).[17]

Both Didier and Annie Anzieu identify the hysteric with a gleaming skin surface, a "double envelope (the child's united to that of the mother) [that] is brilliant, ideal" (AMP 149) or an extra "envelope of excitation" which "not only characterizes the skin-ego of hysteria but forms the hysterical basis common to all neuroses" (249). Annie Anzieu elaborates: "hysterics presents themselves as a sort of excitable surface whose content does not respond to excitation" (AEP 114); often the product of a depressed mother, the hysteric inherits from her distracted care an excess of external and internal stimuli which s/he cannot integrate into the still immature skin-ego. In a recent study, Christopher Bollas associates the hysteric's disturbing and often "toxic" charm (Bollas 139; see also Segal 2002b) with a mother who "unconsciously distributes the child's erotism over the surface of the body, radiating in intensity away from the genital" (Bollas 49). This rather peculiar complaint blames hysteria on a mother who has not sexualized her child in a genitally focused way, so that it becomes unable to engage in "deep sexuality" (51); in general, he asserts, "[the hysteric's] body becomes a genitally decentred erotic vehicle" (62). Whatever the specifics of Diana's childhood or sexual practice, these readings share a belief that the 'gleaming' surface of the hysteric bespeaks a deviation from what Annie Anzieu calls the normal woman's "psychic cavity" (AA 41). Essentially, this is a view incapable of taking on the radical gender implications of the skin-ego. As chapter 8 in this book will explore, sexuality is always a question of surfaces.

As for the question of Diana's being a narcissist or a hysteric, this is the most common remark voiced about her by both psychoanalysts and other people. And it seems a reasonably accurate diagnosis, if diagnosis we want. What it does not do is explain the fascination that the gleaming surface both provokes and represents. To explore this further we need to return to the textual accounts.

All this light seems to have circulated around that "fabulous" (McKnight 38) skin which bulimia remarkably did not affect – "My skin never suffered from it [...] When you think of all the acid!", she marvels (Morton 61) – but the camera always did. In addition to lighting up the space she moved in, she is also a reflector whose exceptional ability was to throw back (throw up) beams from artificial sources. If she was "the unusually multi-faceted reflector of a fragmented and fractious time" (Unsigned Editorial, *The Times*: 25), it is because her beauty, once she had "stepped gingerly into the spotlight" and then "got to the footlights" (Redburn, n.p.), turned from something theatrical to an accessory of the camera: it "leapt through the lenses"(Appleyard: 6) in an "effect of bursting flashbulbs and the dazzle of halogen"(unsigned Editorial, *The Guardian*: 6).

Such mechanized reflexivity might be understood in two main ways. Commonly, but mistakenly I believe in this context, we find the notion that Diana was so apt a mirror because she was herself – at first, at least – a blank or empty reflector. Thus to Hugo Young, unlike John Kennedy who was "the leader of the western world", Diana "was an empty vessel" (Young 19); or to Nicci Gerrard she was "the perfect vessel for our desires" because during the years of her lonely marriage, "her cosseted surface bloomed and her abandoned inner life dwindled" (Gerrard 24). By this reasoning the surface represents the inanition of a proper "psychic cavity": "like a fur coat, a beautiful but empty skin" waiting to "[grow] some insides" (Grant 8). This is the image of an ideal anorexia, the body as a sheer surface, containing nothing. I want to argue, on the contrary, that we ought to understand Diana's skin in terms of the bulimic circuit, an image of exchange in which fluidity is the key. What flows into and out of the dazzling surface of a beautiful woman is gender.

Observe the following four accounts by two men and two women:

> It's funny but when I met her I could swear I could tell she had come into the room even though my back was turned. The first thing that struck me was her glamour. She had the most beautiful skin. The other thing was that she seemed genuinely interested. She said different things to different people. She wasn't one of those important people who goes around saying the same thing to everyone. She tailored the conversation to suit. She was introduced to ambassadors, authors and journalists and every man in the place was turned to jelly. But she was also very funny – using humour against herself. (Lincoln 17)

> In spite of the glamour, she exuded a vulnerability which was puzzling, if compelling. And those who saw her in the flesh fell for her charisma as I did after hearing her address – controversially – a Catholic, pro-family rally shortly after her separation [...] Diana astonished everybody by giving a gentle but extremely liberal speech. (Coward 19)

> And I could see easily how people could fall for her. She was a tremendous natural flirt, and once [...] I was able to imagine myself becoming her victim. She was presenting the Literary Review's annual poetry prize at the Café Royal. I was seated at a table about as far away from her as it was possible to be, but I had the spooky impression that she was looking at me all the time. She wasn't, obviously. But she had that quality which the subjects of old-fashioned oil paintings often have: of seeming to be following you with their eyes, wherever you happen to be. (Chancellor 7)

> I wasn't the first person in my family to encounter Diana. My mother shook her hand many years ago, back in the days when she was still Shy Di, long before she began grooming herself to become the Queen of Hearts. Diana came to my home town in South Wales to open a new children's wing at the hospital where my mother works. I remember my mother arriving home, gushing about how lovely and sweet and kind and caring Princess Di really was. And I remember my reaction – totally unimpressed.
>
> How times change. In my own defence I have to say that the Diana I encountered at the London Lighthouse last year was a different model. [...] I remember thinking how thin her legs were, and how clipped her voice was. It was only when she finished speaking, and got down to shaking hands, that the real value of her visit became apparent. The Di effect was remarkable to behold. Faces flushed with excitement. Eyes lit up. And, corny as it sounds, a feeling of genuine love pervaded the room. For the next three days, I went around telling everyone how lovely and sweet and kind and caring Princess Di really was. (Burston 12)

All these spooky impressions by all these surprised converts share one thing: they are seduced. "She was one of the half-dozen most seductive people I have ever met", announces Vicki Woods: "'seductive, yes. Not sexual', said a former equerry. She flirted with men, women, children". And men, women and children, gay and straight, black and white, old and young – it appears – found in her what they wanted a woman to be: beautiful without being sexual, beautiful and sexual, funny, gentle but liberal, looking at me all the time, shy, groomed, in control, lovely, sweet, kind and caring. This is not the effect of an empty vessel but of a surface in and out of which femininity courses with perfect fluidity.

For the fluidity of gender is, as we have seen, more precisely a fluidity of the feminine. We have noted how the varying valence of body fluids, from the 'pure' to the viscous, from idealized sperm to dangerous blood, depends on its position or movement in and out of place and its greater or lesser coefficient of femininity. Most feared, of course, is its circuit through bodies marked as male. Conversely, it has its sanctioned movement around the female body in such forms as bulimia and radiance. We cannot be surprised to see its politics enshrined in a circularity of the gaze.

All that light – in, out of, around and through this woman's body. In the early days, everyone noted that shy – or was it "coy – upward glance," (Campbell-Johnston: 17) as the cameras slid through her diaphanous skirt or as, more peculiarly, she gazed up adoringly at her royal beau. Peculiarly because everyone knew that she was tall and he was short. We 'don't do it Di' feminists made much of those absurdly posed shots of her carefully placed a step or two below him, or seated while he leaned avuncularly down. The relation of power to the gaze is analysed by Foucault in *Surveiller et punir* [*Discipline and Punish*], where he describes the people looking up to the monarch: "Traditionally, power was what was seen, shown and manifested [...]. Up to this point it had been the role of political ceremony to be the occasion for the excessive yet regulated manifestation of power" (Foucault 219-20). At such moments royalty was on display and the people were allowed to look, not on the face of power, certainly not into its eyes, but at a proper distance and logically from below. Genet's *Le Balcon* [*The Balcony*] (1956) satirizes this relation of mass to icon when the denizens of his brothel present themselves as the Queen, the Judge, the Bishop and the General on the balcony that marks the liminal point between two worlds. The balcony and the television screen are such transmitting skins, two-sided in their function of presenting and protecting (on the double meaning of the term 'screen' [*écran*], see AP 139; also Connor 2001: 36-51 and Connor 2004; 68). We gazed up and, by a certain distortion, we saw her gazing up too. By that distortion, and by the choreographed voyeurism of the royal kiss, we understood certain things: that we were going to be, as Bea Campbell put it with characteristic brutal sympathy, able to "watch her deflowering", and that what we got to see would never be the 'real thing'.

In fact Diana had an "upward glance" because she was, as conventional gender arrangements have it, too tall. This characteristic took some time to be 'grown' to her advantage – "she was very wise, very tall and very beautiful. She loved me", said Arnaud Wambo, aged 8, of Cameroon (cited by Alderson 19) – not only because it conflicted with the sincere objective of being a good wife, but also because it upset the micro-image of monarch and subject that her marriage represented. When she accepted her height – "getting comfortable in this skin" (Morton 46), vertically as it were – she came to embody a strange simultaneity of the dual verticality of power.

Power is vertical firstly, as we have just seen, because the few are on display to the many. Foucault goes on to explain how the individualization that follows from being gazed upon made a major shift in the late eighteenth century. Typically in feudal regimes (of which the British monarchy is a late version writ small), rituals and ceremonies ensured that those with power and privilege become known to their public by an 'ascending' individualization. Over the last two centuries, on the other hand, the downward gaze of a punitive surveillance or discipline individualizes the common man or woman "by comparative measures referring to the 'norm' rather than by genealogies using ancestors as reference points; by 'gaps' rather than deeds" (Foucault 226).

It is in this sense that Princess Diana was, as endless accounts from all quarters marvel, 'one of us'. She was looked upon in a disciplinary mode, full of 'gaps' by dint of her labile femininity and other weaknesses, stared down by the royal family who judged her unstable and more complicatedly by us who wanted to see "the part that wasn't the Princess" (Ruby Wax cited by Johnson: XIV) exposed through the part that was. We also – as we discovered with contrition after she died – must have wanted the discipline by which photographers persecuted her, hounded her out of doors and forced her indoors, with the threat of "face rape", "hosing her down, "whacking her" or "blitzing her" (see Alter 41 and MacDonald 18; the first term is Diana's own, the others are photographers').

Individualized by the disciplining gaze as a "basket case" (Diana's term for the Windsors' view of her: see the Panorama interview) or simply a pretty woman, but also by the ceremonial positioning of her triumphal public role, Diana was a double-facing skin between the feudal and the modern modes of the exercise of power. In this sense too she stands for a circuit in which she was positioned at the middle point, absorbing and reflecting the two modes of gazing. This was most particularly her function for women. She could be adored but also pitied, because whatever misfortunes we think we have endured by virtue of our sex she seemed to have experienced too. We looked simultaneously up and down, as she did. Our lives and fantasies (including our longing to be looked on) were embodied in her.

Thus she was presented for both our gazing and our gawping, in that "gilded cage with a mirror but no bell – and no mate", to quote Bea Campbell again. At the same time, used as we are now to zoos in which the screen is constituted by our car window rather than the ball-and-chain square-bashing of the curious beasts, we were getting in closer. What is skin for, fabulous, comfortable or tortured, but for touching?

The skin consensually connects and contains all the senses but its prime faculty is of course the sense of touch. Although touching never gets us back inside the common skin with the mother, as sexual pleasure and curative miracles tend to suggest, it gets us as near as may be. And so, for every image of light paraded in the literature of the Diana effect, we have two or three images of touch. Liminal again in its value, this image seems to offer some deft combination of the royal touch and the common touch – two notions that are not very different after all, since only the great and mighty can have them. With the 'common touch' (the term is from Kipling's 'If') go all the protestations of her 'ordinariness'. Of course she was ordinary – we all are. Of course she was unique and extraordinary – we are all that too. (We are even all radiant, probably, in the eyes of someone or other.) But she was extraordinarily ordinary, "one and a million" (8-year-old Amiyna cited in Mars-Jones 17), in a way only the great and mighty get to be: God first, and so on down. This explains the ubiquity of her smiling and glowing gaze, perceptible across a crowded room or even through the back of Sarah Lincoln's head, and it also explains the imagery of the miraculous that is typical of 'the royal touch'.

It began with Henry II in England and Guntram, a sixth-century king, in France (see Bloch 21-23). At first, apparently, the miraculous healing was exercised on any disease, but gradually it focused on scrofula, a name for a range of disfiguring skin conditions. It is amazing how often – considering how rarely she must actually have encountered it – leprosy, that archetypal object of disciplinary examination (see Foucault 231-33), is represented among those conditions that Princess Diana was prepared to touch. At its most grotesque, here is a caption in *Hello!*: "She made a point of reaching out to those whom others were loathe to touch, including AIDS sufferers, lepers, and these Indian Untouchables (above)" (Unsigned, *Hello* 15). But the same way with the leprous is noted by a number of broadsheet writers, together with her graceful tactility with other sufferers from visible, mythically contagious conditions. We should not forget that those supposed abject may not always wish to be touched by those wishing to bring them comfort – as a character in Sartre's *Le Diable et le bon Dieu* [*The Devil and the Good Lord*], groaning at the approach of yet another martyr intent on giving him the "leper's kiss", might prove (Sartre 1951: 128). But we do admire and probably envy the gesture by which Diana, often *in camera* rather than in shot, brought her beauty within range of those whom other people find ugly or threatening, whose stigmatized skin marks them out as archetypically condemned to separation.

Of course it could not be literally as much of a "two-way thing" as it seemed and, more importantly, felt (Rudd 17). A close friend of Diana's in her last weeks, Rosa Monckton, reveals that though the princess "had a unique ability to spot the broken-hearted and could zero in on them, excluding all hangers-on and spectators" (Monckton, 'My friend': 11), she regretted the loss of the HRH title because it "did present a barrier, an invisible one, but a barrier nevertheless, which made people stand back a little and enabled her to get on with what she was doing" (Monckton, 'She found': 3). The skin that glowed and "pressed the flesh" (Wilson 11) – as she, presumably, would never have put it – "ungloved" (Moore 1) was also knowingly an armour, even if perhaps not "a skin as thick as an armadillo" (Madonna in Franks: 17). So – did she "break through the carapace of fame [by tearing] at her own scar tissue" (Jenkins 'the young': 18)? Or was she, like the too-beautiful Hélène in *War and Peace* "varnished by thousands of eyes that have caressed her form" (Tolstoy 501)? Or something between the two, held together by a finally cohered skin in which she could feel *bien dans sa peau* (see Mower)?

Certain hints suggest that Diana could only begin the work of touching after she had finished with her eating disorder. In the Morton confessions of 1991-92, she says both "I think the bulimia actually woke me up" and "it's like being born again" since she managed to keep her food down (61). Thereafter she controlled what she brought up – "seven years of pent-up anger" expended in an outburst to Camilla creating a state in which "the old jealousy and anger [was] still swilling around, but it wasn't so deathly as before" (63) – and what she kept back, 'digesting' the public praises that she earlier vomited out. Other hints, by contrast, both in the more guarded narrative of the 1995 *Panorama* interview and in the accounts by Rosa Monckton of her experiences in 1997, suggest that, on the contrary, the strain of servicing her public's emotional demands tended to drain her and led directly to eating binges. How often did she fill herself with food in a bulimic bout, Bashir asked her:

> It depends on the pressures. If I'd been on what I call an 'away-day', I'd come home feeling pretty empty, because my engagements at that time were to do with people dying, people who were very sick, people with marriage problems, and I'd come home and it would be very difficult to think how to comfort myself, having been comforting lots of other people, so it would be a regular pattern to jump into the fridge.
>
> It was a symptom of what was going on in my marriage. I was crying out for help, but giving the wrong signals, and people were using my bulimia as a coat on a hanger, they decided that was the problem. Diana was unstable...

'Anxiety between the four walls' of her marriage and the two of them trying to 'keep everything together' in creating a coherent face for the public prevents the woman, in this scenario, from exploiting an ideal circuit of intimate attention between herself and her husband; instead, a much larger circuit of income and expenditure is set up with the anonymous multitudes who 'cry out' to her. The almost comic comparison between two forms of comfort (for the importance of this word, see Jenkins 'The young': 18), one embodied in merciful acts, the other found in the fridge to be ingested with hysterical haste, shows just how unreliable such hydraulics may be. Diana was "drained and exhausted" (Monckton 'She found': 3) by those who found in the "essence of herself [that she] handed round" a source of "nourishment" (Purves 24) or "sweets" that they could "never get enough of" (Unsigned Editorial, *Independent on Sunday*: 22). Put like this, the picture seems to be of a public in love with her, in the sense that we may all be in love with and endlessly demanding of the good mother. But remember Woolf's Mrs Ramsay, another source of radiant nurturance, who is a lighthouse for everyone else but, left alone, shrinks into vague internal darkness? Because behind the logic of the circulation of gendered fluids around a woman's skin there is not only the positive (bulimic) corollary of liquid excess flowing out of control but also the negative (anorexic) one of dryness, petrifaction, in which all that remains under the garment is a skeletal coat-hanger. We do not, in other words, want the feminine fluid to drain out of our ideal woman. If she were to become skin and bone, what hope would there be for the rest of us?

If this did not happen to Diana and cannot now happen to her, it appears nevertheless with some frequency as the shadowed underside of all that glowing light, occasionally set up as a contrast with the rest of us – for Jan Morris, the more she "floated" as a sort of ideal child the more our ordinary lives were "drab and everlastingly flinty" (Morris 52) – but much more often characterizing the rest of the royal family as that dinosaur dynasty into which she was flowingly inserted and which could not swallow her. Almost all the portrayals of them at the sorry hour of her death paraded the adjectives "stuffy" and "stiff"; where she is fluent, pliable or tactile, or turning "her huge mascara-etched blue eyes soulfully upwards to the camera's lens," they were caricatured as the figure she was not, "standing bolt upright with her handbag on her arm, extending a stiff gloved hand" (these two phrases: Strong 13). Her ready tears made Prince Charles "dry-eyed and stiff" (Showalter 15), "the body held together, the hands that don't touch" (Hall 25); her light exposes the monarchy "to withering scrutiny" (Unsigned editorial, *The Independent on Sunday*: 22) from which they emerge, at the worst, as a "dumb, numb dinosaur, lumbering along in a world of its own, gorged sick on arrogance and ignorance" (Burchill 14), or even "frozen sticks of fossilised shit" (Grant 8).

Why such vituperation, couched in such aggressively desiccating terms? The overriding image seems to be of a dead tree, all armature and no sap. It was less than thirty years, however, since the Queen started that whole process by inviting the cameras onto her estates to film a set of nicely posed domestic groups, giving rise to the 'knit yourself a royal family' metaphor through which one of my teachers expressed a righteous republicanism. These were rationally and quite sensibly rationed breaches of the enclosure of invisibility, and followed logically upon the upward gaze by which we children bought savings stamps in the 1950s with the head of Prince Charles or Princess Anne on them: five Annes equalled one Charles, if I remember aright. Any dynastic system, however rigid and closed, must open up some entry points for the absorption of daughters-in-law. Some digestive juices are needed if the latter are not to be spewed out almost as fast as they are ingested. But can the tall and graceful be consumed by the short and charmless? This seems to be what the question of Diana's beauty is asking.

There are, I would like to suggest, four versions of female beauty embodied in succession by Princess Diana and tracing another cycle of emptinesses and fullnesses. The first is the virgin. Of course she had to be a virgin, as Sarah Ferguson, for instance, did not. As she herself puts it, "I knew somehow that I had to keep myself very tidy for whatever was coming my way" (Morton 28). Who among the rest of us did not watch with embarrassed fascination as the last of the pure maidens was offered up to what we recognized as a merciless machine? Because the idea of an ideally unbreached female body was already thoroughly anachronistic – in other words, a fairytale – and because it seemed as if everybody but she knew the system wanted her as its reproductive conduit: needed her to be simultaneously closed, open and empty.

The anachronism of the royal family, into which the absurd virgin is inserted as its inner tube, thus appears as having no lubrication and no real point of entry. It has to take in but cannot absorb anything not of its own matter; it is immune. Either we did not yet realize that in the 1980s or, if we did, we somehow thought we would not see the process of chewing and spitting-out enacted with the same logical visibility as that weird balcony kiss. What also developed at that time, and increased exponentially in the few weeks after she died, was our not very logical wish to believe her undigestible because she was too much like us – *pace* the fact that the

Spencers are an even older dynasty than the Windsors and offer consistent recent evidence of throwing up and out their incoming women.

The second version of female beauty is the waif. Goodbye Norma Jean, farewell Emma Bovary. What we did not yet recognize as a saint in the making started visibly suffering on the bourgeois-realist scale: the hopeless marriage, the woman frizzled by halogen, wasting away under our gaze. She confesses adultery and bulimia and audiences are at once touched by her eagerness to expose herself and repelled by her apparent skill (manipulativeness). For she isn't only a waif – whether because she is a poor little rich girl, or because she is also a wronged wife (whoever thinks of Monroe as a wife?), or because she shows steely anger, or simply because she is always better to look at than the rest of the family, bending rather than breaking... At this point we ceased to be amazed by their ordinariness any more, sensing that they couldn't do it like she could. Wasn't it interesting how much less shocked we were in *Spitting Image* by the voice of the Thatcher puppet (that of a man) than by the Queen Mother's (the accent of a Northern biddy)? But we no longer derived pleasure from their much less impressive ordinariness, because hers had glamour.

Her charitable and even flamboyant charisma gave us the exhibition we want. We radiated back the pleasure her glowing surface offered us. She had stopped being bulimic by this stage and was dazzlingly transmitting our desires back to us in the form of love. Charm is, according to Camus's Clamence, "a way of making people say yes to you without having to ask them any question" (Camus 62). What she represented – like a present-day therapist, scornful of the couch – was a listening gaze.

Even before she was dead, her daily beauty, like Othello's, set about making them ugly. Most specifically we were both horrified and pleased that Camilla, who had been preferred to her by a man incapable of seeing what we saw, lacked everything she so splendidly had. Diana provided the body that dynasty demanded but the system in which she was essential could not contain her for the peculiar reason that he did not desire her and everybody else did. Her skin implied their shell, her glowing surface their dry matt unmirroring surface. As if the public realized suddenly that what it required of royalty was not simply to be allowed to look at something either paraded or veiled, but to go on staring at something beautiful, something that gazed back beautifully. For this, of course, she soon had to be dead.

"Just before she died", said my mother, "you could see in the photographs that she was happy – no, that's not the right word: fulfilled". My mother was not the only one to notice a difference. Kim Craxton from North Devon observed in true bereavement: "I've got OK magazine, they've got the last pictures of her on holiday and she looks so happy. Thank God we've got those pictures" (caption from 'Nation learns': 12). The fulfilled woman is my third version of female beauty. At last her healthy skin was full of good material, not gorged but loved just right, no more nor less than a person needs. Never mind if he was a foreigner, a notorious playboy, the son of an intruder; he made the English rose feel good. Curiously, the two of them, in the only whispers to have reached us before disaster struck, both use the opposite of the expected metaphor. While Diana reportedly phoned Cindy Crawford to say: "For the first time in my life I can say that I'm truly happy. Dodi is a fantastic man. He covers me with attention and care", Dodi confided in Max Clifford: "'She fills up my senses' [...], quoting a song by John Denver" (Honigsbaum 15). And this may be why we believe them. "Whatever love is" – to quote Prince Charles when he almost certainly knew very well – it is probably a state where each party feels full of the other's good fluids.

The last version of female beauty starts to come out the other side of gender – immortal invisible. Dead, she has no body: the only one who can be with this body is not the king but the king of kings. Her beauty becomes that of the angel. In awe of this new status, Bernie Taupin excised all the lines of 'Candle in the Wind' that might offensively remind us of the waif she used to be, "Never knowing who to cling to when the rain set in" giving place to the completely senseless "Never fading with the sunset..." Diana did not instantly become a saint, for all the selectiveness of memory. She first transubstantiated into that impossible being, a woman with no body. Not just collecting and carrying our sins and suffering like a mix between Jesus and the scapegoat – but also the dustless space on the Louvre wall that people went to stare at when the Mona Lisa was taken down for cleaning. The essential point about that massive pilgrimage (whether spontaneous or choreographed, the precise balance of these motives being certainly unguessable) is that for the first time they knew they were going to see - nothing. That is surely why the grief was both bewildered and free, and why, also, it is not unjust to call it "virtual grief" (Levy 33) or the culmination of a relationship with an "imaginary friend" (Mars-Jones 17), because surely that is what bereavement is, and even friendship in a way. Judith Williamson observes: "just like a message in a bottle, they were outpourings of emotion to someone *who wasn't there*. [...] How much easier it is to pour out all that intensity to someone we didn't know, *who's not there* and, in a sense, never was" (Williamson 8). Letting it all out in that excessive way signifies that the circuit was broken and there was nowhere for the stuff to go except flowing into and through the streets of London. The common touch gets translated into a co-presence of people in one place at one time who are willing, in the absence of any other function, to represent commonality in their own massed bodies. After the funeral and cortège had passed, they hardly knew what to do. "I hope we'll keep in touch every 6 September" (Channel 4 documentary on the funeral), said two women touched by the whole experience.

In the years since these events and accounts in 1997, things have, as they say, moved on. The 'Diana industry' has continued, with books, blogs, supposed revelations and a series of investigations. The first of the latter, conducted in France by Judge Hervé Stéphan immediately after the event, produced a 6,800-page document delivered to the prosecutor in February 1999: it blamed the speed of the driver Henri Paul and his intake of alcohol and prescription drugs. Mohamed Al Fayed continued to call for further enquiries and an inquest was announced for 2003 because "British law insists an inquest must happen when a body is returned to Britain following a death abroad" (news.bbc.co.uk/1/hi/uk/2035106.stm). In the meantime three press photographers were brought to trial in 2003 under France's privacy laws; they were acquitted on the grounds that "both Princess Diana and Dodi Al Fayed had actively courted publicity at times, with their friends sometimes tipping off the press as to their whereabouts" (news.bbc.co.uk/1/hi/world/Europe/3246718.stm); Al Fayed's appeal was thrown out in September 2004. Earlier the same year, the Metropolitan

Police Commissioner Sir John Stevens had begun an inquiry, pledging to look into every detail of both the events and the conspiracy theories surrounding them. It was not the first time anyone said "'We have got to draw a line'" (news.bbc.co.uk/1/hi/uk/3658853.stm) and it was not to be the last. In December 2006, Stevens published his report: the double death was "a tragic accident" (news.bbc.co.uk/1/hi/uk/6179275.stm), not a conspiracy, but Mohamed Al Fayed swore: "'Whatever it's going to cost me, if it costs me the last penny in my purse, I'm not going to rest until I get the gangsters". After two changes of royal coroner, the inquest finally took place, running from September 2007 to April 2008 and concluding that Diana and Dodi had been "unlawfully killed" by the drunk-driver Henri Paul and the paparazzi. The princes thanked the court for their work, Al Fayed departed in despair and the public declared that the £10 million cost had been wasted. The BBC news website of 7 April 2008 quotes 'Mel, England' as saying: "I hope for the sake of her family, this can finally be laid to rest" (news.bbc.co.uk/1/hi/7328754.stm).

During the same eleven years, Diana's children have grown up, less harried than her by the publicity machine, and Charles married Camilla in April 2005, thirty-five years after they first met and twenty-four after his wedding to Diana. A well-wisher, Vivienne Reay from Staines, said: "'I think it's really nice that they're married, they've waited long enough'" (news.bbc.co.uk/1/hi/uk/4428161.stm). A drama called *Charles and Camilla: Whatever Love Means* and portraying their relationship in a positive light was released the same year on television and later on DVD. A major influence on the mellowing of the public's attitude to the royal family has been Stephen Frears' universally well received *The Queen* (2006), in which Helen Mirren and Michael Sheen played the Queen and Tony Blair with remarkable sensitivity and warmth. On the other hand, on the eighth anniversary of the crash, members of the public are quoted by the BBC as saying it was especially important to lay flowers at Kensington Palace "'because of the wedding'" (news.bbc.co.uk/1/hi/uk/4201022.stm); and the most cursory internet trawl still finds comments about ritual murder and evergreen memories alongside vituperation against the royal family and advertisements for grief counselling.

The tenth anniversary of Diana's death August 2007 was marked by a flurry of events. Channel 5, for example, ran compilations with titles like *Charles and Diana: The Wedding* and *Diana and the Camera*. *The Times* reissued sections from its 1997 editions. A thanksgiving service was held in the Guards' Chapel at Wellington Barracks, near Buckingham Palace, chosen by William and Harry because it was "the closest thing to a mother church for their regiment, the Household Cavalry" (Unsigned 2007: 7). The media were most interested this time in who would not be there: Camilla, Paul Burrell, the princes' girlfriends and the general public. The Rev Richard Chartres, the Bishop of London, who gave the address, said: "'It's easy to lose the real person in the image, to insist that all is darkness or all is light'" (Chartres 5). The accompanying article in *The Times* concludes: "Princes William and Harry hope that yesterday's service will draw a line under commemorations of their mother, and that the public memory of her will fade to a warm glow" (Hamilton 4). In among the resting in peace and lines being drawn, this is about the only remaining mention of Diana's radiance.

Chapter 6: The surface of things

In the last two chapters, two famous bodies were considered as variously representative of a gendered circuit of desire. This chapter and the next look at forms of the relation between a desiring or enquiring subject and its others. In the next chapter the object will be a human body conceived as a skin-enwrapped object which the subject seeks to enter and be. In this one the other is, intrinsically at least, an ungendered object, a thing. How do we reach our hand or our eye, those most desiring of organs, towards something which is not ourselves, which seems to make a demand on us, and yet which does not reciprocate or give way to that demand? Nietzsche wrote in an aphorism of 1880: "Things are only the borders of people" (Nietzsche 49). By this he surely means that they mark the outer point of the territory that human bodies imagine as their controlled space - the edge of what Paul Schilder was later to call 'the body image' (see also Weiss), where objects cease to be prosthetic and are perceived as absolutely other. What does it mean for a desired thing to seduce or disturb us by its solidity, its existence as a 'full' object having a surface but, in a sense, no inside? Outsides appear to speak of insides; indeed, it has often been suggested that the whole history of science is a protracted wish to explore the resisting inside of objects. What might it mean to us when it turns out to be... nothing?

In this chapter most of my referents are poems and essays in which a protagonist reaches towards, or feels repelled or frustrated by, an object - but these acts of reaching or repulsion are verbal ones. Writers are writing texts about objects. Some of these verbal things are figurative, others affect the protagonist as though they were. What common verbal space do the human and the non-human occupy, and can this common occupation in any sense be consensual?

Texts act upon us in ways that, precisely because they are not material, may imitate and thus seem to replace the way material things act; like fetishes, they both do and don't supplant the thing they are not. Barthes is surely referring to this seeming seductiveness when he writes: "the text is a fetish object, and *this fetish desires me*" (Barthes 45). How is the surface of things represented in poetry? The possibility of using words to represent the very quality they do not possess - solidity - is one that preoccupied certain poets of the late nineteenth and early twentieth centuries; it is one of many ways in which language is made, under the art-for-art's-sake principle, to do the strictly impossible. What I am interested in here is how surface and mass are used in a non-dimensional medium to conscript the motion of desire. Texts seduce, and in doing so they invoke the movement of eye or hand towards something one ought to be able to caress. These poems are concerned with the skin-like interface between language and things, alive and dead (imitating in words the similarity of 'still life' and *nature morte*), perishability and permanence. Is the artist's attitude to the imagined medium of stone that of Medusa or that of Pygmalion: flesh turned to stone or stone to flesh? I shall be looking first at texts by Conrad Ferdinand Meyer (1825-1898), Georg Büchner (1813-1837), Théophile Gautier (1811-1872) and Rainer Maria Rilke (1875-1926).

Meyer is best known as a composer of *Dinggedichte*, or 'thing-poems', poetry that reproduces objects in words – often, though not always, culturally mediated objects such as paintings or sculptures. His 'Der römische Brunnen' [The Roman fountain] (1882), representing a fountain in the Borghese Gardens that Rilke also took for one of his *Dinggedichte*, imitates not only the form of the fountain but also the fluid relation between two media, water and stone, and two states, motion and stillness. This idea reappears developed into an aesthetic in the poem 'Michelangelo und seine Statuen' [Michelangelo and his statues]. In four of Michelangelo's statues, a bodily action is suspended: the slave does not groan, the thinker does not feel his heavy helmet, Moses does not leap up and Mary's tear does not fall. Each material image "represents the gesture of suffering [...] without suffering" (Meyer 63), offering humanity the representation of torment and even death overcome. The densest of media seems the only proper place for this occurrence: "what tortures living flesh" is "rendered blissful and delightful [*beseligt und ergötzt*] in stone". The poet like the sculptor will die, but stone will not. The most curious thing about this argument is that it celebrates the least 'solid' of the attributes of a sculpture – its function as a snapshot.

There is a moment in Büchner's novella fragment 'Lenz' (1839) when the protagonist imagines a similar process of 'snapping' a human group in order to 'preserve' it in stone. It is, he reflects, a murderous, medusan act:

> As I was walking up the slope of that valley yesterday, I saw two girls sitting on a stone; one was fastening her hair, the other was helping her; her golden hair hung down, and she had such a serious pale face, though still so young, with her black outfit, and the other girl busying herself so carefully. The finest, most intimate pictures of the old German school hardly come near it. Sometimes one would like to be a Medusa, turn a group like that into stone and call people to come and look at it. They stood up, the lovely group was spoilt; but as they climbed down, between the rocks, I could make out another picture. The most beautiful pictures, the fullest tones, gather together and loosen apart. What remains is the one thing, beauty: an unending beauty that moves from one form into another [...]. You have to love the human race if you want get inside the special individuality of each person, no one must seem too lowly or ugly; that is the only way to understand them. (Büchner 234-35)

Meyer and Büchner take up opposite political positions on the question of sculpture counteracting the perishability of human flesh: the first regrets that it is impossible, the second celebrates the fact. In Meyer's poems, language cites the virtues of the materiality of massive things in order to do what language actually does better: stream and rest at the same time or capture the instant that in three dimensions would never hold still – for texts only exist in the communicative encounter that has no duration because it is always virtual.

Gautier, who declared himself "a man for whom the outside world exists" (Goncourt 343) and argued – partially ironically – in the Preface to *Mademoiselle de Maupin* (1835-36) that "everything useful is ugly" (Gautier 1973: 54), published a collection a few months before his death in 1872 which he named after the most tiny decorative objects, enamels and cameos, *Émaux et camées*. I want to look first at his version of the impulse we have seen described by Meyer and Büchner to freeze flesh into the solidity of art. Fascinated by the surface of beautiful women, he returns repeatedly to a medusan, anti-pygmalionesque fantasy, turning flesh to stone in a kind of necrophilia.

In 'Le poème de la femme' [The poem of woman], subtitled 'A Paros marble', a woman "reads [...] the poem of her beautiful body" to the "dreamer who loves her" (Gautier 1947: 7); after a series of erotic tableaux she finally throws her head back, rolls her eyes and, mimicking orgasm, "dies of ecstasy" [*volupté*] (9), leaving the poet to kneel and pray beside her lovely corpse. Another poem contemplates a plaster-cast of the hand of a beautiful lady 'Impéria', the "pure fragment of a human work of art" (11). The strangest of all the woman-to-stone poems is 'Coquetterie posthume' [Posthumous coquetry], the testament of a female narrator ordering how she is to be arranged in her coffin in her flounciest dress, among pearl-strewn pillows, and made up with rouge and kohl so that, we imagine, the man who has left her will want her once it is too late. In all these ways, the transformation of flesh into an art object is meant to make the imagined woman – with familiar *carpe diem* tones – service a fantasy of frozen compliance. Another set of poems plays literally with the shades of surface, following a spectrum from pink to white. In 'La Rose-thé' [The tea-rose], "the most delicate of roses" (90) vies with the seventeen-year-old skin of 'Madame'; in 'A une robe rose' [To a pink dress] the lady is, like her dress, a structure of surfaces, textures so fine and so similar that it is impossible to tell where silk ends and skin begins.

These poems are not just exquisite gallantries. They embody as linguistic matter the horror of a desire that dwells, as all desire must, on the surface. Surfaces are made of tissue whether fabric or flesh: and their tints are on a spectrum from the flush of living things to the stillness of dead white. There is, as Büchner's reading of art's medusan rage shows, a cruelty in this move that is none the less violent for being so cold. The coldness of creativity fascinated Gautier as much as it did his admirer Mallarmé – for instance in the latter's poem 'Le vierge, le vivace et le bel aujourd'hui', another *Dinggedicht*, in which the "unflown flights" of writing are the virtual, impossible motion of a swan trapped in ice.

There is, however, a contrary movement in Gautier's poetry, one that changes from a superficial aesthetic and inverts the medusan impulse of petrifying flesh into a frozen state into something more violent. Surface here no longer offers a canvas for consensual play but invites a wound: the object becomes the containing thing asking for penetration. In the closing poem, 'L'art', art alone survives the perishability of matter because of a particular way in which it contains:

> Sculpte, lime, ciselle ;
> Que ton rêve flottant
>   Se scelle
> Dans le bloc résistant !

> Sculpt, file, chisel;
> let your drifting dream
> seal itself
> in the resistant block!

The 'drift' of life can only be captured by being incised into a resistant medium: something without substance has appeared 'through' the massive bulk of a thing that contains it – a thing that can only contain if it is penetrated by an artist. What fantasy is this?

    To answer this question we need to look at a poem in which Gautier pulls together all the arts, high and low, in a gallery of objects that evoke what to him is the quintessential aesthetic encounter, the desire of a double-sexed body. The poem is called 'Contralto', but it begins by describing the fascination of a statue of Hermaphrodite in the Louvre. The beauty of this "enigmatic statue" (30) with its "malicious pose" – it reclines with its back to you, looks like a perfect woman but tiny male genitals are revealed if you move to the other side – is described in three epithets: "disturbing" [*inquiétante*], "accursed" [*maudite*] and "multiple". Such an object is monstrous, a chimera, the supreme combination "of art and sensuality", and most importantly, it invites the approaching gaze to penetrate it:

> Bien qu'on défende ton approche,
> Sous la draperie aux plis droits
> Dont le bout à ton pied s'accroche,
> Mes yeux ont plongé bien des fois.

> Though it is forbidden to come near you,
> under the drapery with its straight folds
> whose end is caught on your foot,
> how often my eyes have plunged!

They haven't, though, of course, because they can't. Teasing for exactly this reason, the statue provokes because it cannot be penetrated or 'found out' (for a similar conceit in reference to a Bonnard painting, see Serres 27). There is no answer to the question whether it is Aphrodite or Cupid, because it has to be both. It is, after all, in its very massiveness, nothing but an uninterpretable surface. Gautier goes on in the rest of the poem, to compare this beauty to that of the contralto, "the hermaphrodite of the voice", to the combination of Romeo and Juliet, chatelaine and page, two fluttering butterflies, melody and harmony, Cinderella and her friend the cricket, Zerlina and Mazetto, Kaled and Lara, and the commingled sighs of a man and a woman making love.

The ending of 'L'art' represents the fantasy of being and having: the "dream" flies in or is already in, or has to be cut in. Containment is the aesthetic act completed. But for the poet of 'Contralto' the aesthetic response is not one of contemplation of a thing contained, it is desire, a movement-towards, and the object of desire must be enigmatic; you want to penetrate but cannot because the doubleness must be impossible to 'unclothe' – otherwise, solved, it is no longer beautiful. In this argument, the beautiful object is a tease, the aesthetic impulse tends to penetration, and its frustration is the essence of what remains.

In his epistolary novel *Mademoiselle de Maupin*, Gautier follows this fantasy into prose. The heroine is a woman who dresses in male costume in order to see and hear (and ultimately feel) what men think about women when no woman is present. Desired by two lovers, D'Albert and Rosette, she enjoys being adored but is by definition unable to undress. Finally she lets herself be found out in a nicely crafted production of *As You Like it*, playing Rosalind in the revelation scene, and ends the text spending a night with both in turn. Just before this momentous conclusion, she writes:

> My dream would be to have both sexes by turns, in order to satisfy [my] double nature: today a man, tomorrow a woman, I would save for my lovers my languorous caresses, my submissive, devoted gestures, my most abandoned caresses, my melancholy little sighs, everything woman-like, cat-like in my character; and with my mistresses I would be enterprising, bold, passionate, with a triumphant air, my hat over one ear, and the swagger of a swashbuckler or an adventurer. Thus my whole nature would come forth and be visible, and I would be perfectly happy, for it is true happiness to be able to develop oneself in every direction and to be all that one can be.
> (Gautier 1973: 394)

Like the voice of 'Coquetterie posthume', the figure of Madeleine fulfils a male fantasy, but she goes beyond it into a different one: Madeleine does not die, but leaves her two beloveds behind so that she can pursue further adventures, enjoining them to "love each other in memory of me, whom you have both equally loved, and sometimes say my name to each other in a kiss" (416). This departure is the only logical end for her because, like the poet and object of 'Contralto', she is unwilling to choose or be chosen: either transitively or intransitively, desire cannot undress.

The aesthetic fascination with surface in Gautier's writing is always libidinized. It is sexual in a way that remains disturbing, accursed and multiple because it takes a particular angle on the Medusa fantasy set up by Meyer or Büchner. As the human mind wishes to exert creative control on the flux of material existence, the spectrum is always from pink to white, living to dead, flesh to marble. These desires are very close, even though they invert them, to those of Pygmalion. Surface is understood as that which does not, but *should* yield. Otherness, even that of objects, is a sexual scandal. Beauty is the surface where both sexual ambiguity and the ambivalence of flesh and stone reside – and tease.

I want to move on now to a different representation of the surface which, in sculpture, attracts and disciplines desire. Rilke is also the author of many

Dinggedichte, but his objects are represented very differently from those of either Meyer or Gautier. They do not mimic solidity but invert it. Thus a tower is described by the noun "Earth-inside" [*Erd-inneres*] or Saint Sebastian is "like one lying down" [*Wie en Liegender*] (RNG 58 and 35). 'Das Rosen-innere' [The inside of roses] begins: "Where does this inner-ness have its outside?" (143), and the exquisite 'Der Ball' [The ball] enacts in a single apostrophic sentence the circular trajectory of being thrown, falling and being caught. In this thing-poem the object is absent, replaced by an 'idea sealed' in language; only its attributes, its combination of weight and weightlessness, the arc of its limited flight, its precipitation by players who wait for its gravitational return, exist in the seventeen lines which thus trace a circle of loss and desire. Characteristic of the German language, there are no Latin roots, every complex idea is made out of simple ones, there is a sort of chemistry of notions that allows the materiality of things to emerge out of the ways in which they are impalpable: an object becomes a form without either substance or surface. What occurs here, then, by contrast to the poetry of Meyer or Gautier, is that language is restored to its position of superseding, not imitating, the massivity of stone.[18]

When Rilke writes two studies of the work of Rodin in 1903 and 1907, he moves from an impulse to undermine materiality to a preoccupation with it, but again it is not solidity that interests him. The surface of the sculptures is essential, not because it is a fantasized entry point but because it is not. The surface of the human body represented in stone "consists of countless meetings of light with the thing [...] there were endless places and in none of them was nothing happening. There was no emptiness [...] There were only numberless living surfaces" (RR 309-10). A face – Rilke gives as an example Rodin's early *The man with the broken nose* – "has no symmetrical surfaces, nothing that repeats itself, so that no place remains empty, silent or indifferent" (317); and beauty is not intrinsic but "arises out of the feeling of equilibrium, the balance of all these moving surfaces with each other, and the realization that all these moments of agitation swing out to their end inside the thing itself" (318). Just as "there is nothing in nature but movement" (319), so in an art-object the aim is to present "the restlessness of living surfaces" (320).

Discussing the 'conditions' [*Bedingungen*] under which a beautiful art-object might come to be, Rilke writes:

> Whoever pursues these conditions attentively to their conclusion will discover that they go nowhere beyond the surface, nowhere inside the thing; and that all anyone can do is fabricate a surface that is closed in a particular way, with no place on it subject to chance, a surface that, like that of natural things, is surrounded, shaded and lit by the atmosphere – only this surface, nothing else. [...] Nothing exists except a single, thousand-fold moving and inflected surface. (381)

The importance of gesture here is neither the capture of life's changefulness nor the nobility of representative meaning, but something that will seem to emerge from an inner necessity onto an outer screen; Rilke likens its visibility to what is "waiting enclosed in a hard bud". The relation of the outside to the inside is shown in the gesture of the *Walking man*, the pregnant *Eve* or the *Inner voice* which all seem to be "listening in to [their] own body" (325). As Rilke's contemporary Hugo von Hofmannsthal (1874-1929) puts it in an aphorism: "Nothing exists in the inside which is not simultaneously perceptible on the outside" (Hofmannsthal 40). Hofmannsthal is interested, far more than Rilke, in the 'inside' of a thing, but he too sees surface as its essential form: "The depth must be hidden. Where? On the surface" (47; see also Valéry 215: "The most profound thing in man is his skin").

Most radically of all, Rilke offers an image that changes the whole schema from the bilateral one of human/object desire that I have called medusan or pygmalionesque to something on a quite different scale:

> when Rodin endeavoured to draw the air as close as possible to the surface of his things, it is as though he dissolved the stone into it just there: the marble seems to be simply the firm, fruitful kernel and its final, lightest contour is swinging air. (RR 369)

Here the 'skin' of a sculpture is not its own surface, to which our hand might reach or into which our eye might try to penetrate, but the virtual line drawn around it by light, space, air – and the orbiting human subject. This is a quite different kind of 'common skin', an aesthetic of consensuality that takes surface further away from the edge of things, towards more nucleic circuit of desire.

As we saw earlier, Rilke's thing-poems turn solidity into something more like aery form, and his conception of sculpture is as "a single, thousand-fold moving and inflected surface" standing as the nucleus of a prokaryotic cell. I want to turn now to a different kind of object, one that, to all appearances, should command the most tender human response. That object is the doll.

Rilke's essay on dolls has two significant sources. One is his own experience in childhood. Born as the replacement child for a dead sister, he was named René ['reborn'] and dressed in girls' clothing longer than was normal even at the time: a pair of photos on facing pages in a biographical album show him aged five and seven uncannily mirrored as a girl and a boy (Schnack 46 and 47). He is said to have hated being passed around among his mother's friends like a big doll (Peters 41; Freedman 9). At the age of ten, he was sent away to the military academy of Sankt-Pölten, vividly memorialized by Musil in *Die Verwirrungen des Zöglings Törleß* [*Young Törless*] (1906). Like his protagonist Malte Laurids Brigge who as a child presents himself to his mother as little Sophie (he actually has a dead sister called Ingeborg) and discusses what is to be done "that Malte" (RMLB 94-95 and see Segal 1981), Rilke is not sure whether to wish his boy-self away or what to do with the girl-self who seems so much nearer to the mother's desire. I shall return later to the significance of this position as what Anzieu calls "the living dead" in relation to the lost sibling.

The second influence is the 1810 essay by Heinrich von Kleist (1777-1811), 'Über das Marionettentheater' [On the puppet theatre], in which the interlocutor argues startlingly that the movements of puppets are more graceful than those of human dancers. The point, he says, is not to think of the operator manipulating a multitude of individual strings but to see motion as arising out of the weight of the figure:

> Every movement, he said, has a centre of gravity; it is enough to control this in the middle of the figure; the limbs, which are merely pendulums, follow mechanically by themselves, without any need of further action.
>
> He added that this movement is very simple: whenever this centre of gravity is moved in a *straight line*, the limbs describe *curves*, and often, if it is shaken in a purely random way, the whole figure will fall into a sort of rhythmic movement like a dance. (Kleist 556; see also Parry, Bergson [1940] 1981: 23-25 and 52-61, and Paska 410-30)

The puppeteer does not need to be a dancer – but in a mysterious way the very manipulation of the puppet means that, even though the operation is quite simple, its line "is nothing other than *the path taken by the soul of the dancer*; and he doubted whether it can be found unless the operator transposes himself into the puppet's centre of gravity, in other words, unless he himself *dances*" (Kleist 557; see also Diderot [1773] 1951: 1035: "a great actor is [...] a marvellous puppet whose strings are held by the poet").

It is like the art of the maker of sophisticated prosthetic limbs: both simply require an understanding of "proportion, flexibility and lightness – all in the highest degree; and especially a more natural disposal of the centres of gravity" (558). Puppets, unlike living dancers, never behave affectedly and they never resist the natural grace of submitting to the gravitational force. No human being can equal the almost weightless quality of puppets in which "the force that raises them into the air is greater than the one that binds them to the ground" (559). Kleist concludes that "it is quite inconceivable that any human being could approach the gracefulness of the puppet. Only a god could equal inanimate matter in this regard; and this is the point where the two ends of the circular world meet and grasp each other" (560).

In Rilke's Fourth Duino Elegy, the angels, whom we met in chapter 5 "drawing their own streamed-forth beauty back into their own countenance", appear in a similar encounter with puppets. It takes an angel, entering the "brat" of the puppet's physical husk, to "yank it up. Angel and puppet: that at last would be drama" (56-57). We humans stop this drama from occurring, incapable of being either pure spirit or pure matter. The nearest thing we might be capable of is the martyrdom of the *saltimbanques*, puppet-like tumbling and leaping on an "unutterable mat" (line 95).

The German word 'Puppe' means pupa, puppet and doll. In each of these senses there is the idea of the incompleteness of the figure as a husk or shell whose content is a virtual soul, supplied from elsewhere. Kleist - who alternately uses 'Puppe' and 'Marionette' - shocks us by arguing that puppets *almost* have a soul of their own, which the puppeteer only supplements by giving in, like them, to the forces of gravity and flotation. Rilke's essay is most definitely on dolls, not puppets, though his title refers not to his own childhood toys but to decadent gauze-clad figures created by Lotte Pritzel which he saw in a Munich exhibition in 1913. Though they have something in common with the automata featured in Hoffmann's 'Der Sandmann' (1815), the ballet *Coppelia* (1870), and surrealist fantasies like those of Hans Bellmer, Pritzel's dolls were not life-sized, and it is the question of childhood that concerns Rilke. In often obscure terms, he expresses a long-standing dislike or fear of dolls. They are described as "false fruits whose germs never come to rest, sometimes almost washed away by tears, sometimes exposed to the flaming aridity of rage or the desolation of neglect" (RP 265), by turns desired and despised. Dolls are present at events like birthdays or illnesses, accompanying the tenor of strong emotions, but unlike dogs for example, they somehow "take no trouble" over these things, and by night "they *let themselves* be dreamed, just as by day they were tirelessly inhabited by alien forces" (RP 267).

Like dogs, *things* are far more cooperative. To them "even the roughest wear and tear is like a caress" which may diminish them but at the same time gives them a "heart" that fills their material form and makes them as touchingly mortal as we are. This harks back to the second essay on Rodin (1907), which takes the form of a lecture and begins by asking members of the audience to recall "any one of your childhood things" and to recollect whether anything was closer to you, more necessary and trusted, than such a thing. Wasn't everything else except that thing capable of hurting you or being unfair, frightening you with an unexpected pain or bewildering you with an uncertainty? If goodness was one of your earliest experiences, and confidence and not feeling alone – don't you have that thing to thank for it? Was it not a thing with which you first shared your little heart like a piece of bread that had to last for two? (RR 377)

Such everyday objects willingly played any role: "animal or tree, king or child – and when it stepped back that was all still there" (378) – and they laid the ground for more complex experiences: "it was with this thing, its existence, its particular appearance, its final falling to pieces or mysterious vanishing-away, that you experienced everything human, even deep into death" (378).

In 'Puppen' (1914), Rilke lists such objects as "a needlework stand, a spinning-wheel, a domestic loom, a bride's glove, a cup, the binding and pages of a bible; not to mention the mighty will of a hammer, the devotion of a violin, the good-natured zeal of horn-rimmed spectacles" (RP 268). The 'heart' of these objects lies surely in their utilitarian functions – they are good servants – but he has something more subtle in mind when he contrasts them with dolls, for the difference lies in the ways in which the two kinds of objects receive our imaginative input. Things somehow respond; dolls do not. As children we tried to love them but in reality we always unconsciously hated them; despite all the "purest warmth" (269) we lavished on them, they lay there fatly like a "gruesome foreign body". Unlike people, dogs or the things cited in the Rodin lecture, a doll never comes half-way to meet imaginative play: "it gave no response, so we had to take over the whole work, splitting our gradually expanding character into part and counterpart, and to some extent using it to keep at bay the boundless world that was over-flowing into us". So, whereas the thing was capable of expanding and sharing our "little heart", the doll splits and exploits it, returning desire with silence.

A doll, says Rilke, maintains an absolute silence not because (despite its affected appearance) it gives itself airs or thinks itself sublimely unresponsive like God but simply "because it is made out of good-for-nothing, totally irresponsible matter" (271). The key insult here is an external appearance that seems to promise human responsiveness but, because it disappoints, proves humanity itself a sham. The problem is (again) the relation of outside to inside in a figurative object. "A poet could fall into the power of a puppet [*Marionette*], for a puppet has nothing but imagination. A doll has none, and it is exactly as much less than a thing as a puppet is more" (272). In the range of aesthetic possibilities, then, the angel and puppet, standing at either end, carry fullness and openness in their most creative forms; between them, human and doll are derisory substitutes, and the encounter between them cannot succeed. It dominates the child because it is neither person nor thing, and because it appears to hold out something unimaginable and longed-for: "the doll's soul" (273).

In a final and particularly bitter comparison, immediately following this ringing epithet, Rilke opposes dolls to a variety of masculine playthings: rocking-horses, toy trams, a ball, a tin trumpet. They have clear-cut, inspiring souls; but the girlish, ghoulish "doll-soul" is insinuating, respects no boundaries:

> one could never quite say *where* you were. Were you in us or in that sleepy creature over there, whom we constantly talked you into? What is certain is that we often relied on each other and in the end you were not in either of us and got trodden underfoot. (274-75)

There is something irritatingly fluid in the doll-soul; it has a demonic aspect – "not created by God" (276) – and we cannot quite detach ourselves from it; like an unworn garment it is prey to moths, and the more we try to cultivate it in the doll the more the larvae of those moths devour it from the inside, and then flutter forth to fling themselves into flames with a burning whose "momentary smell floods us with boundless unknown sensations" (277). In the doubly flowing end-point of this metaphor of insides-turned-outsides there is a feminine uncanny that the doll revives in the adult male Rilke.

This image of the moth-ridden, flaming garment, the feminine uncanny, resembles nothing so much as the haunting of Marguerite Anzieu as described by her son:

> So my mother was conceived as a replacement for the dead child. And since she was another girl, they gave her the same name, Marguerite. The living dead, in a way... It's no coincidence that my mother spent her life finding ways to escape from the flames of hell... It was a way of accepting her fate, a tragic fate. My mother only spoke openly of this once. But I knew it as a family legend. I think her depression goes back to this untenable position. (APP 20)

And while for Marguerite the uncanny is partly the effect of the same-sex, same-name replacement to which she is subject, for Rilke it is focused on the particular gender demands of the other-sex replacement. Though this is not explicit, we might surely see the sister in the doll-soul he is made to incorporate in response to the mother's bereavement. How is he to manipulate that displaced self once the muslin dress is exchanged for a sailor-suit or cadet's uniform? The aesthetic response, the wish to remake infancy through shaping a verbal world, leads to the puppet, the angel and the art-thing. In the Fifth Elegy, the *saltimbanques* give soul to their action as a midway between puppet and angel; but the doll is too painful to insert into poetry: like Gide's use of Madeleine (inside and outside his fiction), Rilke's dolls are the projection of his mother's "threatened and threatening" bereavement into a creature he has to repudiate.

Unlike Rilke, the relationship with toys that Gide describes from his childhood is scientific rather than imaginative: he tells us of a marble he painstakingly extracted from a hole in the wall or a kaleidoscope he opened up and refashioned (GJ3 113-14 and 83). Is it always the 'soul' of toys that we seek when we take apart a mechanical toy or dismember Barbie (see

Bignell 36-47, or the playful destruction of a doll in Gottfried Keller's 1855 'Romeo und Julia auf dem Dorfe')? There must be, we feel, some ghost in the machine of a plaything, especially a figurative one. In identifying the disturbing way we 'insert' ourselves into the husk of a toy, Rilke is highlighting the seductive demand that objects seem to make upon human beings, all the more upon those whose psychic content is still labile and immature. In Condillac's gradually animated statue (1754), or in Gide's representation of the education of the blind girl Gertrude as the eventual emergence of a soul when her teacher sees a smile appear on her "statue-like face" (GR 889), we find the same pedagogic fantasy as he entertains towards his young lover Marc Allégret, hating his rival Cocteau for making him feel like "Pygmalion finding his statue spoiled [abîmée], his work of art destroyed; all my work, the trouble I'd taken as an educator, my ideas completely ruined" (Pierre-Quint 391).

Puppets and dolls both seem to promise a dream of retroactive animation – which will only lead to grief. A differently gendered version can be found in the story of Pinocchio, first published by Carlo Collodi in 1883. Even though he clearly has no strings, Pinocchio is always described as a puppet – the Italian is *burattino*, a glove-puppet; this is perhaps because the more usual term for what he is (*bambola*, a doll) would carry connotations of femininity not only in his potential user but also in him. The same ambivalence is visible in the doll-like, man-like function of the nutcracker in the eponymous ballet. In Rilke's preference of rocking-horses or toy trams we find a similar yearning to grow, through the infantile love of figures, out of an uncannily feminized stage into boyhood and thereafter manhood. The desire of Pygmalion for Galatea or Gide for Marc is more overtly sexual, but gender questions seem endemic to the relation of human and thing.

We are fazed by the otherness of objects. We need – and fear – the object that is not flesh. Do we create statues in order to undo the radical difference of the inanimate? An automaton like Cesare carries out the fantasies of Dr Caligari or the hero Francis or the *auteur* Robert Wiene, just as Olimpia represents those of Coppelius, Nathanael and Hoffmann. The use made of these uncanny figures is not so very different from that of the transitional object – half 'me' and half 'not-me' – in which Winnicott sees the first moves of an infant's ability to play and later to create, to manage its world. All the examples I have examined in this chapter are encounters between men and the imagined residue of material things, which go dangerously in and out of the actual hardness of their created status. This play is less like the standard relationship of a child with its transitional object than like that of an autistic child, whose preferred object is hard-edged. Frances Tustin describes how autistic children, terrified both of the excessive flux of self and other and by the shock of leaving the container of the mother's body, supplement their "protective autistic shell" (Tustin 19) with "idiosyncratic inanimate objects" (29) like a Dinky car, the buckle of a belt, even a cannonball taken to bed (95, 103, 115). "The main purpose of autistic objects [...] is to shut out menaces which threaten bodily attack and ultimate annihilation. *Hardness* helps the soft and vulnerable child to feel safe in a world which seems fraught with unspeakable dangers" (115). It is allied, of course, to the phantasmatic creation of the "armour skin" for which Anzieu cites Tustin and Bick – but it is more directly a negative creative act, one that, as Tustin describes it, provides a barrier rather than a bridge to reality: "nothing can get in but, more important still, nothing can get out" (118).

If, as researchers are currently arguing, the development of autism is related to the quantity of testosterone found in a foetus, then autism may indeed be an extreme form of masculinity (see Knickmeyer *et al* 1-13 and Baron-Cohen 245-54). The risk, though, is that hard autistic objects are "liable to snap and break" (Tustin 121), and that, since they are more like "fetishistic objects" (125) than everyday transitional objects, they will also fail to provide the covering effect of the half-delusion typical of the fetish.

We return herewith to the opening position of this chapter. How does an aesthetic object come to seem a fetish "that desires me"? In his history of sculpture, Alex Potts traces how figuration in concrete form has developed since the neo-classicism of the eighteenth and nineteenth centuries, through radically anti-figural phases, to a contemporary reintegration of the body idea in a range of new media; he highlights "aspects of modern engagements with the sculptural that have to do with the physical, sensual and affective dimensions of the encounter between viewer and work" (PSI 4). The location of a sculpture, whether crowded into a Sculpture Garden at the turn of the twentieth century, coolly presented in the 'white cube' galleries of more recent times, or displayed out of doors among city- or landscapes, is always "both a private and a public affair" (21). We meet the object or installation with both our body and our imagination; and perhaps because, in a postmodern communicative world, we no longer believe in an "authentic, intensely private communion with the sculptural object", we are confronted rather than receptive.

I want to look at two aspects of this history, first the idea of sculpture as sexual object and second the question of solidity and surface. To begin with the neo-classical nude, idealized female figures such as those of Canova or Pradier combined the ostensible "representation of a human subjectivity at one with itself" with "an obvious erotic charge" (16). In Herder's assertion that men are more responsive to a female sculpture and women to a male one, Potts identifies "a very un-Greek gendering of viewing" (33) – by which I understand him to mean rather a heterosexualizing of viewing. We have seen in Gautier's 'Contralto' the move from the anti-Pygmalionism of woman-to-stone to the more radical idea that sculptural eroticism consists in the surface's hints of a double-sexed charge. This is, he implies, the general logic of figurality taken to its most perverse and thus most aesthetic point. Whether altogether heterosexualizing or not,[19] this position is masculine, not simply because we have been looking, throughout this chapter, at male viewers, but because, whatever the sex of the viewing body, the stance of this desire is penetrative. That it attributes seductiveness to the object is also the normalizing position of masculinity. Canova's figures may "seem aware of being looked at and admired" (57) while those of Thorwaldsen "are staged so as to block any explicit awareness", but this is consistent with the alternative 'provocative' or 'sideways' looks in pornographic images, two versions of the standard 'come-on' of the feminized object.

The erotic appears differently in Minimalist work, whose physical effects are less representational than visceral: of his "apparently totally unsensual large-scale plinths, cubes and beams" (250), Robert Morris said in 1975: "'Their sensuality has to do with their shape, how they stand in space [...] I just think that there are certain shapes that one [...] gets [a] charge from'" (ellipses Potts'). On a 1968 work by Donald Judd in which stainless steel cubes mounted on a wall are lined with vivid yellow or amber plexiglass,

Potts comments: "What sharply separates the conception of Judd's work from that of a Greek nude is the categorical erasure of any direct representation of desirable forms or body images in the work itself" (305); rather, the 'sexiness' effect is produced by "the internalized body image induced in the viewer by being in the presence of the work". The erotic sense of body presence is, thus, displaced from representation to something else: "the viewer is precluded from projecting its form as the image or objectified visual model of the subjectivity it intimates" (306), and the experience is effectively an introjection rather than a projection. By this I mean not a part object psychically absorbed as in the Kleinian term, but a direct sense of encountering something of the inside-body, one's own or another's, one's own through another's.

This introjection effect appears also in Potts' closing sections in relation to three women artists. Eva Hesse's *Laocoon* parodies the subject of Lessing's famous essay by consisting of a ladder bound in paper mâché on which cloth-covered cord falls in "drooping, twisted snake-like forms" (343) in a witty "undoing of heroic virility" that imitates both the snakes in the first-century statue and "vital bits of its inside, the internal writhings as well as the source of pain". Similarly in Rachel Whiteread's inversion of large-space objects – a bath, a house – into solid casts, Potts finds a "slight frustration", "low-level irritation" (360-61) which I would venture to identify also with a certain sexiness, irritating to some, gratifying to others, that has to do with the 'insideness' of the sense of body they convey. And Louise Bourgeois's *Cells*, enclosed spaces, enterable for the viewer by eye (partially) but not with the whole body, stage "dramas of confrontation" (362) because here there is an inside, physical and psychic, at which we can only peep.

What, then, does sculpture have to tell us about our own fantasies of solidity and space? All sculptural things dramatize the confrontation between the body of the viewer located in a specific place and the 'body' of the object s/he finds there. As we have seen in Rilke's analogy of the sculpture with the kernel of the fruit, this confrontation could be understood as taking place inside a 'common skin'. The "swinging air" that

Rilke describes surrounds both bodies but one breathes it in and out while the other does not, and for the first of these, the ability to move, circling round the sculpture, is part of a sense experience midway between touching and seeing. If the sculptural object is different from other objects by the 'staging' of its location, the viewer is as much located *by* it as freely consenting to respond to it. Potts's history looks at the changing relation of response and staging. Chary of evaluative terms, his favourite word for what distinguishes a sculpture is 'compelling'. This term suggests a stronger than usual demand to look. Where this demand is felt to reside is, of course, a pointless question, but it seems to belong to a certain indeterminacy of the magnetic relation (the 'sexiness' we have just looked at) and a certain ambivalence about the solidity of form. A viewer of Minimalist works, for instance, finds an "indeterminate compulsion to continue looking intently [because] there is always 'something simultaneously approaching and receding'" (PSI 196; the internal quotation is from Fried 1967); or, "the unusually direct engagement with presence and solidity and materiality (in works by Carl Andre) is compelling because it involves a negation as well as a reactivation of the sculptural" (311).

Solidity, the supposed serene autonomy of the ideal object, typically characterized figural sculpture, both classical and (less unselfconsciously, goes the theory) neo-classical. By a variety of techniques, the dazzle of over-white surfaces was tempered by yellow or rose tinctures or even soot (Körner 216); but, despite this effort to resemble flesh more than stone, Canova's sculptures "strike one as being without apparent depth" (PSI 98). In contrast, as we have seen, Rodin's human figures, with their "swellings and lumps", seem to embody human depth accompanied by an "almost deathly calm associated with this" (99). The particular imperative of modernism was "to undo the massiveness of sculpture [...] one is made most intensely aware of volumes, of enclosures and openings, of the blocking out and division of space" (303-04). One is also made aware of one's "own body's occupancy of space – a sense of being not an inert solid mass weighed down on the ground, but an arena of interiority extending outwards into its immediate surroundings while also being sharply delimited and clearly located in one place" (304). The 'skin' of an object changes its meaning when "the surfaces strike one more as an activated boundary between interior and exterior than an unyielding outer face of an inert solid lump" (305). We see here the development of Rilke's theory of the kernel. What does a 'thing' then become?

In a 1968 interview, Carl Andre defined a thing as "a hole in a thing it is not" (312). This recalls the marvellous logic of Gregory Corso's "What are fish? Animalized water" or André Breton's perverse self-image as a "soluble fish" (Corso *geocities* and Breton 55). In all three fantasies, solidity is the sudden aberration of another medium – liquid or gas, in either case more fluid than it. All objects would be mistakes, miraculous, dangerous or comical, whether they are too much *there* or too little: we want things to stand clear against other things – but not too clear, for either excess seems to wound our tender flesh.

If objects, and especially sculptural objects, stand in space, traversed on their surface or gaps by air which they are not, an extreme version of this is the *Marsyas* of Anish Kapoor, which filled the huge emptiness of Tate Modern's Turbine Hall in 2002 with an arrangement of voluminous funnels made from 3,500 metres of sweeping red plastic-coated material tensely bound to three circles made of forty tonnes of high-strength structural steel. The title cites Titian's disturbing *The Flaying of Marsyas* (1575-76), but more specifically it refers to the skin-like effect of the membrane viewed from inside or out.

Kapoor's earlier work had always been "anti-vertical. My sculpture seems to have a downward energy" (Kapoor 60). He had frequently used red before, "trying to turn the red of earth and body into sky" (62), and to upturn mass into volume: "the space contained in an object must be bigger than the object which contains it" (13). The enormous scale of this object is both intimidating and at the same time "intimate. [...] Human presence is always implied" (62). It must feel like a thing "we know, but somehow don't know we know" (61) – in other words, our body. Walking through, around and underneath this huge construction, the viewer ends up on the central bridge, above which a funnel of stretched membrane hovers a few feet up, containing focused light, and the effect is to feel "pulled" or "sucked" in (63) by this "very sexual, emotive kind of form" (60).

As described by Kapoor himself and by his colleague, structural engineer Cecil Balmond, the funnel-shape is like an ear, a throat, a mouth, but also like a range of other orifices; Kapoor recalls "a totemic image of a man wearing another man's skin" (61) and observes: "I'm interested in the unknown here. There's one real unknown, rather like this man wearing the other man's skin – it's as if he's got a ring around his mouth, two little rings round the nostril holes, a ring round each of the eyes, and a ring round the ears. It's as if those holes define the intermediary volume" (63). We are thus both entering the unknown of another's body (wrapped itself in a different person's skin) and already there: the known/unknown of the body is both our own individual space and, uncannily unfamiliar. The *Marsyas* image is of suffering – "a symbol of the transformation that occurs in the crucifixion" (61) – but also of the homeliness of being in a skin. Balmond comments: "the skin seemed to grow flesh beneath the red lacquer. I could see the technical abstractions we had struggled with – of warp, cutting patterns, and tension trajectories – materialize and thicken into muscle and bone and sinew. [...] The imagination is exposed – raw, vibrant" (69). The "visceral effect" (62) of the funnels creates a kind of "vertigo", an experience of falling that is disorientatingly inverted: "falling doesn't always have to be downwards. The falling in some curious way can also be towards a horizon or even upwards" (62-63).

Insideness, or what I have called the 'introjection effect', is central to this piece. The way that the sculpture blurs "the two-dimensionality, the surfaceness of the thing and the very three-dimensionality of it" (64) is a combination of the funnels drawing the viewer in and the sense of encountering a skin:

> Skin is a consistent quantity in everything I've ever done. It's a notion that I've talked about in my work for twenty years now. Skin is the moment that separates a thing from its environment, it is also the surface on which or through which we read an object, it's the moment in which the two-dimensional world meets the three-dimensional world. Seemingly obvious statements, but I think that looked at in any detail they reveal a whole other process. There's a kind of implied unreality about skin which I think is wonderful. (64)

For Anzieu, the myth of Marsyas, flayed for competing with the music-making of Apollo, is "a coding of the particular psychic reality that I call the Skin-ego" (AMP 68). First it links the "envelope of sound" to the "tactile envelope (provided by the skin) and secondly it stages the reversal of a maleficent fate (inscribed in the flayed skin) into a beneficent fate (the conserved skin preserves the resurrection of God, the maintenance of life and the return of fertility to the land)" (AMP 68-69). The "unreality of skin", in Kapoor's modern Marsyas, is simultaneously that of a man wearing another man's skin like the replacement child and the inverted space of those funnels that both engulf and hover like a maternal body unwilling to come to rest in a common skin: no beneficent fate here. If the surface of sculptures seduce and resist, submitting only provisionally to our wish to touch, enter or surround them, what happens when we attempt to touch, enter or surround the body of another human being?

Chapter 7: In the skin of the other

In the last chapter we saw Anish Kapoor create his enormous *Marsyas* out of the fantasy of "a totemic image of a man wearing another man's skin". This chapter will explore the fantasy of being inside the skin of another human being.

When Gide looked at a photo of Pierre Herbart, a handsome young friend of Cocteau's whom he met in 1927, he said "'I really think he has the physique that I would most like to inhabit'" (CPD2 205). We need to distinguish this fantasy of entry inside the other from a notion of sexual penetration. What the latter means we began to explore in the last chapter and will return to in the next. But in the instances I want to look at here, the skin or external appearance of another is not so much the object as the context for desire, the imagined pleasure of being rather than having. This is the desire to live as another person, don their appearance, in order to do something we cannot imagine doing any other way.

Here, for example, is a governess finding herself literally in the shoes of her admired employer:

> A strange thing about those shoes was the way in which, when she was wearing them, Mrs. Brock, who was a heavy treader by nature, planted her feet and walked with the same long steps as Lady Grizel, and stood in the same careless, rather flighty way. A lovely sort of fantasy possessed Mrs. Brock as she moved in this new pretty way, this confident way. Part of herself became Lady Grizel – she absorbed Lady Grizel and breathed her out into the air around herself, and the air around was a far less lonely place in consequence (Keane 20-21).

It is not always such a pleasant fantasy. Flaubert sent Louise Colet a letter in April 1853, in the early stages of writing *Madame Bovary*, where he complained of the feeling that he was being drawn into characters he resented:

> *Saint Antoine* did not cost me a quarter of the intellectual tension that *Bovary* demands. It was an outlet; I had nothing but pleasure in the writing, and the eighteen months I spent in writing its 500 pages were the most deeply voluptuous of my whole life. Consider then, every minute I am having to get under *skins* that are antipathetic to me. (Flaubert 297; see also GJ1 1245 and Segal 1998a: 118-20)

Gratifying authorship, in this image, is an orgasmic outpouring; painful authorship forces Flaubert to look out from inside the skin of hateful characters. I have explored elsewhere what this seems to mean to Flaubert, and how the intense involvement with characters whose despicable nature is to be somewhat like himself creates the particular demands of an aesthetic of 'objectivity' both within and across the gender divide (see Segal 1992: 115-22).

In similar vein, Anzieu cites Starobinski: "'Flaubert represents in the body of Emma sensations he has felt himself; and he feels in his own body the sensations he has represented in the carnal subjectivity of Emma'" (ACO 119). More generally,

> A text is a *chef-d'œuvre* when, out of what his life has left unused and unknown to him, the writer creates a work in which the hyper-reality of evocations and the uncanny familiarity of their consequences gives the reader the feeling of entering a dream or living a hallucination which represents, localized at the margin of his own body, an other part of himself. (ACO 225)

Creativity is one version of this sublimatory impulse; others are less overtly purposeful. And as we shall see in the cinematic examples that follow, the assumption of a false self can prove, like a second skin, difficult to slough off again. Thus Musset's eponymous Lorenzaccio, after years of acting the part of companion in corruption to the duke his cousin whom he wishes to assassinate, recognizes with despair that "vice used to be a garment – now it has become stuck to my skin" (Musset [1834] 1976: 118).

The original purpose that motivated disguise is no longer there 'inside' the gestures and actions he has aped too well - indeed, this mimicry seems to prove that he never can have been the innocent he thought. An act of futile and suicidal murder is, after this realisation, "all that remains of my virtue" (119).

Whether motivated by "virtue", curiosity or a more sinister end, the desire that assumes the costume of another's identity will, like Lorenzaccio's or Mademoiselle de Maupin's, find the garment hard to remove - like the psychical tearing of the original 'common skin'. In the fictions of this chapter, we shall trace three versions of the multiplication of desire through the body of an other. These fictions are Andrew Niccol's *Gattaca* (1997), Anthony Minghella's *The Talented Mr Ripley* (1999) and Spike Jonze's *Being John Malkovich* (1999).[20] In the first two of these, a male figure takes on the bodily existence of another for reasons of combined envy and desire; in the third, two protagonists, male and female, enter the body of John Malkovich, and act out their converging/diverging desires from within it. In all three cases there is an implicit focus on an art we considered in the last chapter: puppetry. When puppets dance, or John Malkovich imitates a puppet dancing - or when Matt Damon or Ethan Hawke pretend to be Jude Law, or for that matter Jude Law pretends to be Dickie Greenleaf or Jerome Eugene Morrow - what 'possesses' what, body or 'soul'? Is the exterior mastered by something contained inside it, or does the container shape and control its contents? Is the relationship between appearance and self as labile as acting, dancing or fantasy seem to make it? Why did the late 1990s have a particular obsession with the exchangeability of skins?

Gattaca is a dystopia set in "the not-too-distant future" (opening titles to the DVD; see Wood 2002: 166-73 and, for a comparison between *Gattaca* and *Memento*, Litch 2002: 141-60). The eponymous space station is spelled with the four letters of DNA, G-A-T-C - which appear also in their dazzling permutations on an identity screen: above the genome, 'Morrow, Jerome', and above the name, in large capitals, the word 'VALID'. The face in the image is a blend of the faces of actors Ethan Hawke and Jude Law. Blended in a way genes cannot be (they fall and reassemble by the mechanics of dimorphic reproduction, a negotiation, not a melding), these facial features stand for the possibility of inside becoming outside. The theme of the film is a study of the forensic use of body elements – inner stuff converted into readable evidence – masquerading as a triumph of masculine aspiration over the limitations of the body. In order to reach the stars, Vincent Freeman has to lodge his 'dreams' in a better body. His ambition needs Jerome Eugene Morrow's frame to function in.

Vincent [Ethan Hawke] is the elder of two brothers. The product of love in the back seat of a car, he is what is known as a 'god-child', despised for being assumed rather than chosen: his genes carry a likely early death, whereas his younger brother has the carefully selected genes that make him a 'valid'. The epithet 'in-valid' follows Vincent everywhere but most especially into Gattaca where, his father has warned him, "the only way you'll see the inside of a spaceship is if you're cleaning it". Cleaning is the work of keeping in order surfaces he wants to penetrate. To enter, you press a finger-tip against a reader that takes a drop of blood and judges you with the buzz of a red or green light. The most intimate stuff from depth or surface – a smear of blood, a drop of urine, a single hair or flake of loose skin – acts as a forensic index of worth, the inner and outer being similarly indices of self as legitimate or illegitimate. If such body bits are read as markers of self, how are they to be controlled in such a way that 'you can', in the terms of the American dream, 'be whoever you want to be'. Inner and outer space is the large context for this, and the story is that of the 'inner self' (soul, love, aspiration, a myth of personal being) becoming outer (physical appearance, the arbitrary limits of birth in this new aristocracy). In order to get off "this ball of dirt" (Eugene's phrase), Vincent has to appear to be someone else, and that someone is Jerome Eugene Morrow [Jude Law]. He has to borrow his body; but does he enter his skin?

In the end, this is the question I think *Gattaca* poses. The beautiful frame of Eugene is bought by the determined Vincent in order to provide him with the space to enter space. Vincent wants to go up in a rocket, and doesn't think beyond the pleasure of being permitted to leave. On the way he achieves three versions of approval: the love of 'the girl' [Uma Thurman]; the assistance of a couple of subordinate father-figures, Caesar and Dr Lamar, and the sacrifice of Director Josef [Gore Vidal], convicted of a murder Vincent/Jerome has been suspected of; and triumph over his brother Anton. Anton is motivated by the knowledge that he is superior, but (for all we see) by little else: we find him in adulthood in the guise of the 'investigator'. Brought in to solve the murder by intricate inspection of bodily forensics, he realizes that a stray eyelash belongs to the in-valid Vincent – but how to convict him, seeing that he is passing himself off as Jerome? We never wholly divine Anton's thoughts – the suggestion is that he wants to save his brother from discovery, though the sting of rivalry is never resolved: he hates the idea of Vincent's triumph – but it may be that he, like the others but without the space-fantasy, chiefly wants knowledge. To whom does the body part evidentially belong? How has the technology been made to lie? Whose face do you see when you compare identity image with fleshly head? In the two main male pairs of this story, we see fraternal difference played out in two ways via the virtual similarity of the body: Vincent and Jerome use the body for an economic transaction in which advantage trades with desire; Vincent and Anton, who have drawn (but not simultaneously or similarly) on the same gene-pool, play out their rivalry by moving in parallel lines. The quintessential, and finely conceived, occasion of this parallel, is a swimming contest they call 'Chicken'.

Present place in this film is always bounded – chiefly by the fathomless sky which Vincent intends to penetrate, but elsewhere, more locally, by the sea. Again the surface of an element is only apparent; it is, more accurately, a context for conquering. So the closest to real passion Vincent and Irene get is to gaze together, in respectful wonder, at the starry sky – and it is after their first sex that the lover rushes out to scrub himself on the seashore. (In the early screenplay, it is a field, not the beach, and the stage direction reads: "There is a haunted, tortured look in his eyes as he tries desperately to rid himself of himself". Falling in love has - via sex, its bodily translation - exposed him to the risk of being found out as 'merely' his own body.) But in childhood, the sea stood for something very specific: competition. Vincent and Anton would arrange footballs and fruit on the beach to imitate the planets, and then swim out side by side to see who would give up first and turn back. The clever thing here is that each must estimate not so much his total available strength (measured against the other) but half the total, for he must be able to reach land again, turning round at the exact midway point to exhaustion, "always knowing each stroke to the horizon was one we were going to have to make back to the shore". Just once, the adolescent Vincent beats Anton, indeed saves his life, seemingly by knowing his own half-capacity just that bit better than his brother, but actually (as is revealed at their last meeting, when they swim again just before Vincent takes off) because "I never saved anything for the swim back". In this hare-and-tortoise morality tale, those gifted by the technology that nature has turned into will inevitably suffer from complacency: sublime recklessness is available only to the truly driven.

Earth-bound, grown-up Anton pursues others, hosts the order of things. He knows that a stray eyelash or skin-flake is not self but evidence. We are used to this principle of matter out of place as the source of delight in such detective series as America's *CSI* or the UK's *Silent Witness* or *Waking the Dead*, in which, in one way, the 'scene" can never be clean enough for the criminal, but never soiled enough for the beautifully groomed and sensitive detectives. They and their prosthetic machinery - red goggles and a green beam expose invisible spots of blood or semen, careful careful tweezers gentle out a single hair from a car-headrest, brute evidence of foregoing violence now turned into seeming order - see what cannot (but must) be seen. There is nowhere to hide from their gaze which is actually not so much penetrating as determinedly superficial, superficial at such a level of expertise that my skin is no longer my home but only the source of bits that will betray me.

Remember that Vincent was destined only to enter a spaceship by being its cleaner? In our contemporary world - from which this fiction is meant to be "not too distant" - hygiene has long since become an affair of pursuing invisible enemies, for which even domestic technology is never good enough, and our meagre housewifely skills quite inadequate (see Hardyment 1988). Recently doctors have started citing the disadvantages in protecting our children from even the most invisible dirt: healthier in some ways, they suffer in others, from unaccumulated resistances or the allergies that thrive in urban spaces, the hovering pollen or lowering fumes. Invisibly awaiting us 'out there' in the air, the enemy is also from us and of us: bad bits fly off to harm ourselves and others. Naked Vincent scrubs himself on the beach after a night of love because his psychic world is that of the psoriatic: every flake of skin must fall, and it falls live from the orbit of his control. So how is he to have the skin he 'deserves'? By purchasing the 'good dirt', the valid evidence, of Eugene Morrow's body. *Gattaca* is a fantasy of the revenge of dirt on cold-hearted hygiene: Vincent/Cinderella will go to the outer-space ball, by dressing up in Eugene's droppings.

Eugene is in a wheelchair since a self-inflicted accident some years earlier. His lower limbs – though even thus he is taller than Vincent, who has to undergo a gruelling operation to lengthen his legs – are useless: we watch him at one point, when Anton is about to find them out, drag himself excruciatingly up the spiral staircase of his elegant condominium. Vincent buys Eugene/Jerome's body because he needs his identity in order, we have seen, to zoom away. But what's in it for Eugene? What I want to look at particularly here are two connected things: what motivates Eugene's part in the other man's metamorphosis, and what precisely is the process by which Vincent turns into – or rather does not turn into, as we shall see, for he both already is, and never will be – his other?

As far as Eugene Morrow's motive is concerned, the story gradually emerges that he despaired of life because he never was – of course – quite perfect enough: he waves a silver medal bitterly in Vincent's face, saying "Jerome Morrow was never meant to be one step down on the podium". He dies in the closing frames still clutching the medal, on which we glimpse the figures of two men swimming side by side. Thus in his own way (inverting the to-and-fro tracks made in the sea by the Vincent/Anton rivalry), he has followed the reverse trajectory to the aspirant: having discovered he could not be the best, he chose to die, threw himself in front of a car, and ended up in a half-life, half-wasted, 'uselessly' handsome – because what is beauty actually for? we shall return to this – smoking and boozing in his chair, alone indoors, watching the sky from a small elevated window beside the incinerator-space, with no other seeming purpose than a monetary one.

Actually gratuitous action for oneself or another who represents oneself seems to be the base of this parable. The process of change takes an unexpected route. In the end, Eugene is left behind, apparently a shell, watching Vincent soar in his space-prosthesis, having made the ultimate tribute: "I got the better end of the deal. I only lent you my body. You lent me your dream". Leaving behind a fridge full of specimens for the return Vincent does not talk about, he consigns himself to the incinerator - hell to the other man's heaven, down to his up. Has he, like the other helper-men, offered himself up for the sake of the hero's triumph? It is true that we expect him to disappear once Vincent goes away, because his function is to supply, not to demand. But I wonder exactly why this is; for they are not hydraulically exchangeable according to the fraternal equation "Is that the only way you can succeed: to see me fail?" Nor do they exactly blend either, despite the identity picture that puzzles the investigator. There are two reasons for this.

First: the process of borrowing body detritus is never complete: Vincent never starts 'growing' Eugene's body parts of his own accord. This is most definitely not a cyborg fantasy. Body and technology, for all the intentions of the civilization the film represents, never do work together; they are always inimical. So Vincent never gets a transplant or transfusion of Eugene, he has to renew the process laboriously day by day - this is *Groundhog Day*, not *The Fly*. It is the laboriousness, the risk (when he throws away his contact lenses to avoid detection, he is almost run over), the hard work, of the process that is the moral basis of the tale: masculine aspiration is no easier than feminine housework, both are reproductive, apparently purposeless repetition, the mountain never conquered, until suddenly it is, and you are no longer Sisyphus/Prometheus but the eagle. The lesson is that you really will only see the inside of a spaceship by cleaning it, and by cleaning yourself, over and over again every day. But then, one is tempted to ask: what is the point of seeing the inside of a spaceship?

Whatever it is that Vincent thinks he is going to (and we shall pursue this fantasy in the next chapter) it is as much a further degree of enclosure as a wide-open vista. In mocking mood, Eugene says: "Anyway, the room I'm stuck in is bigger than your tin-can [...] I go places in my head". And Vincent, "going home", as he lyrically thinks, imagines the object of his aim as "the closest thing to being in the womb". The refusal to think about his return in a year's time suggests that Eugene by dying is imitating rather than serving him.

The second reason takes us into an extra-diegetic area. I want to suggest that it is the non-identity of the two men for the audience that provides part of the effect of the film's myth. Intra-diegetically, the disguise, although always provisional, is successful. "You look so right together I want to double my fee", says black market agent German. "We don't look anything alike", Vincent replies. "It's close enough", says German; "when was the last time anyone looked at a photograph?" Both their eye colour and their accents are different, yet it is understood that with the prosthetic adjustments this will not matter: "He's a foreigner" - "They don't care where you were born: just how". And it doesn't. Bit by bit the identity-swap 'takes'. Typing composedly at her work-station, Irene doesn't connect the old Vincent identity image flashed up on her screen with the 'Jerome' she is about to seduce. When Vincent wants to give up, Eugene retorts angrily: "You still don't understand, do you? When they look at you they don't see you, they only see me".

Forensic evidence confirms only so much - Director Josef, who turns out to be the murderer, argues calmly: "Take another look at my profile, Detective. You won't find a violent bone in my body". What proves identity is who they want to see; and by the end, enough people want to 'know' Vincent is Jerome for them either to believe him or pretend to. Like Eugene, they desire his success because he proves the possibility of 'exceeding potential' that the system is designed to preempt. The question I want to come back to is: do we? In one important scene, Eugene is asked by Vincent to be himself again for the day - "I was never very good at it, remember?" he responds. Confronted by Anton in the presence of Irene, Eugene acts as the Jerome Vincent has since become. Everything about him, accent, eyes, figure, demeanour, differs from the other man. When a blood test brings up the Jerome Morrow identity screen, both the others are shocked. "Who were you expecting?" smiles Eugene.

For his lover, it is important that Vincent can plead "I'm the same person I was yesterday" - though he then turns to Eugene and says half-jokingly: "I think she likes us". For his brother, non-recognition is at first mutual, and the audience is uncertain when or whether each knows who the other 'really' is. The question is here, of course, what 'really' means. *Pace* Lorenzaccio 'true' inner motivation is certainly preserved intact in this film; this is why the labour of disguise needs to be renewed each day, this is what takes Vincent up in the rocket. In *Gattaca*, what makes a self self, the defiance of genetic destiny, is assigned to the mystery of the transcendent will. Will the audience be willing to share that *parti pris*?

Do we as an audience come to accept it as reasonable that Vincent/Ethan Hawke is taken for Eugene/Jude Law? As it happens, this question will arise again in relation to my next film, *The Talented Mr Ripley*, in which again Jude Law plays the body to be assumed. Having acted Lord Alfred Douglas the same year in Brian Gilbert's *Wilde* (1997), Law established himself as the quintessential beautiful Englishman – an Englishman capable of playing Americans as well, as other films show. In *Gattaca* his English accent is preserved: why doesn't Anton remark on it? we surely do. Or rather, why did Andrew Niccol choose an English actor and not ask him to speak in an American accent, if not in order to subject viewers to an extra test that the characters do not undergo? When the two men are first seen side by side, it is not true to say, as the early screenplay did, that "we are struck by the similarity of Eugene's face to Jerome's", and again we must assume that Niccol chose to present us with this challenge. To return to a question that I put in relation to Anton: whose face do you see when you compare identity image with fleshly head? When an audience watches an actor playing a part, what difference does it make that Ethan Hawke has taken over from Jude Law the task of playing a character squatting inside the latter's body?

If, ultimately, we see what we want to know, we virtually invent the blended face of the neo-Jerome Morrow, which neither we nor the characters in the fiction have ever actually perceived outside a photofit. It appears that Eugene/Jerome accepts this logic, for he disposes of himself as the credits roll – down with the rubbish, like a golden Gregor Samsa. Or rather, he knows he has not been blended, and perhaps this is what makes him the most sure of being redundant. The replacement, in this film, both can and cannot replace: the final question is: does the lock of hair in the card carry a quite different message from the urine sacs in the fridge: I have existed?

The obliging Lamar who lets Vincent go through knowing he (like his own son) is an in-valid with the soul of a hero but the urine and the left-handedness of an fallible mortal has, in an early scene, proved that phallic is as phallic does. As Vincent pees, or pretends to, he looks down and comments: "Beautiful piece of equipment [...] an exceptional example". And so, after all, if we needed to be told, it turns out that the only prosthesis a man needs is the one he has already (almost) got, and all he needs to be recognized by is his desire. If we in the audience desire to envy this, we will believe. There is a pleasure in overriding the evidence of our senses – or in other words in suspending disbelief – if it allows us, like Eugene, or for that matter like the two actors, to 'borrow' Vincent's dream. To do that we must declare our body a housewifely inconvenience, a temporary measure.

In *Gattaca*, the problem of desire begins as something between the individual and his body – *his* body, since the story is entirely about masculinity, as we have seen – but it quickly becomes a question of duality: of who owns the man's aspiration and the body needed to enact it, and of how other men may be either rivals or helpers in his individual plot. In Minghella's *The Talented Mr Ripley*, again essentially the story of a single psyche, the issues are triangulated. How Tom Ripley [Matt Damon] will use and destroy others, chiefly the object of aspiration Dickie Greenleaf [Jude Law], in order to be someone he has never had the opportunity to be, is shown not only through a series of splits in the self – stripes across Damon's face in the opening and closing sequences, repeated flashes of body fragments in Italy's ancient sculptures – but also, as Minghella remarks in his commentary to the DVD, through the frustrations of moments disrupted: there is always a knock at the door, a twist, or a meeting that has to be conducted in front of a third party.

The plot is simple: Tom Ripley, a footloose New Yorker, is invited by shipping magnate Herbert Greenleaf to go to Italy to persuade his son to come home. Greenleaf deludedly thinks Tom and Dickie were friends at Princeton; Tom lets him believe this and embraces the unhoped-for opportunity to see Europe. He enters the charmed world of 1950s Italy, where villages gather at water's edge to celebrate the birth of the Madonna and boys on mopeds applaud fanciable girls in the Rome streets. His meal-ticket is charming, feckless Dickie who, with his girlfriend Marge, represents a class born to fine objects and limitless cash changed at American Express offices under the eyes of obliging functionaries. Dickie is beguiled by Tom's talent to impersonate and his artful borrowing of Dickie's own passions: jazz, Italy. But when he tires of him, as Marge has warned he will – "The thing with Dickie: it's like the sun shines on you, and it's glorious; and then he forgets you and it's very very cold; when you have his attention, you feel like you're the only person in the world, that's why everybody loves him so much" – his attachment turns to anger and he murders him. After that it is easy and gratifying to pretend he is Dickie; moving to Rome, he lives in luxury, takes on that classy world – until upper-class Freddie finds him out, and he must be murdered too. Thereafter the talents are turned to the art of evasion: Tom and Dickie by turns, depending whom he is with, brilliantly covering the tracks of both as the police gradually come to suspect Dickie of having first murdered Tom, then Freddie and then committed suicide,[21] he finally faces Marge, who understands everything but is ignored; Dickie's father, who has decided his son is the culprit and is best buried; and Peter Smith-Kingsley, with whom he might have been perfectly happy but whom he is forced in the end to kill as well.

In this summary, I have followed the common plot of the novel and the film, up to the last few lines, in which I take Minghella's conclusion because this is the version I am focusing on here. Both Highsmith's novel and Clément's *Plein Soleil* are, in different ways, studies of a gifted and unlikeable go-getter, a petty criminal who supplements his many lucky breaks with a talent to amuse and an opportunistic ruthlessness. In both cases the idea of killing and impersonating Dickie [Philippe in *Plein Soleil*] precedes the moment of murder and the motive is predominantly practical. Minghella's Ripley is written in a tragic mode, more psychologized, and even though Highsmith's third-person narrative does not prevent us from entering him sympathetically, Minghella requires us to experience through his viewpoint the trajectory that takes him from the basement to the sunlight and back again, ending with the terror of having murdered his true love-object and remaining in darkness for ever. The fractured mess of his inner world begins and ends with a series of chances. He starts out as a Cinderella who "has the manner but not the manner born" (AM commentary), whose native gifts include a sensitivity to classical music which he can play only if a Princeton graduate has a broken wrist or the lights are down at the concert house where he works brushing dandruff off the shoulders of rich men in tuxedos. He ends as Cain, carrying the mark of his crimes in the form of an unconscious loaded with darkness, a locked basement containing himself.

Minghella's triangulation of Tom Ripley's story consists in a number of interpolations in or expansions from the earlier versions. *Plein Soleil* – to which, curiously, Minghella never refers in his commentary – is foreshortened at both beginning and end, consists mainly of scenes on boats, and presents a firmly heterosexual Ripley motivated by financial and sexual envy, who buys himself Philippe's girl once he has taken over Philippe's money. Highsmith's story is more complex sexually but lacks the pathos with which Minghella draws us in.

Sexuality in Highsmith's novel is always to the fore, but in the special sense of a discomfiture in Tom Ripley with his undeclared desire and confused masculinity. Brought up by an aunt who mocked him for effeminacy – "'He's a sissy from the ground up. Just like his father!'" (Highsmith 34 and 86) – he had been "a skinny, snivelling wretch with an eternal cold in the nose" (35) and occasional murderous impulses, until he learned to repress his differences by exploiting his looks, "the world's dullest face, a thoroughly forgettable face with a look of docility that he could not understand, and a look also of vague fright that he had never been able to erase. A real conformist's face, he thought" (31). With this face, his native boredom and the philosophy that "something always turned up [...] He was versatile and the world was wide!" (12 and 32), Tom knows he can go wherever he wants by impersonating other people. But the "vague fright" remains, and it is tied up both with a sense of being pursued, which fulfils itself once he moves up from small-time fraud to real crime, and with being dimly implicated in other people's homophobia – it is after Dickie has reactivated his aunt's taunt and reminded him that "'*Marge thinks you are*'" (86) that Tom decides to kill him and take on his identity. His hatred of Marge for her girl-scout cheeriness and broad hips is little less virulent than his instant dislike of Freddie's bulk, red hair and self-assurance. Excessive gender embarrasses him, and he keeps his own at bay by believing Tom Ripley the negative at the base of the structure. He recalls the humiliation when his witticism "'I can't make up my mind whether I like men or women, so I'm thinking of giving them both up'" (71) was roughly repudiated by a friend, but much later: "as a matter of fact, there was a lot of truth in it, Tom thought. As people went, he was one of the most innocent and clean-minded he had ever known. That was the irony of this situation with Dickie" (for discussions of homosexuality in the film see Street, Keller, Bensoff and Gabbard).

If the motive of murdering Dickie is, in Highsmith, tied up with the enigma of Tom's sexual repression, and in Clément with a simple case of financial and heterosexual rivalry, in Minghella it is complicated by a broader class fetishism: Tom wants what Dickie both is and has. Dickie is clothed in the ease and charm of those born to wealth: his skills (playing the saxophone, swimming, skiing) never seem to have to be rehearsed, whereas we watch Tom practising the identification of Chet Baker or Charlie Parker just as he practises Italian. Once he has killed Dickie, he uses his money to enter that world, whose accessories are the body to be put on: we see him smooth and assured in his new suits, having cast off his glasses like the proverbial Hollywood heroine, using Dickie's voice and wearing his rings. As Highsmith puts it, when he sets out back to Mongibello after killing Dickie:

> Tom had an ecstatic moment when he thought of all the pleasures that lay before him now with Dickie's money, other beds, tables, seas, ships, suitcases, shirts, years of freedom, years of pleasure. Then he turned the light out and put his head down and almost at once fell asleep, happy, content and utterly confident, as he had never been before in his life. (Highsmith 97)

Among Minghella's triangulating adaptations is the addition of two female characters: Meredith Logue and Silvana, rich girl and poor girl, who both love Dickie, though silly Meredith loves him in the guise of Tom. The pregnant Silvana serves to illustrate, by her real suicide as opposed to Dickie's false one, the damage we do as tourists "treading with big heavy feet" (AM commentary) on other people's cultures. She also provides the opportunity to show Dickie being violent when cornered, a difference from the Highsmith original that allows Tom to murder almost in self-defence when he is taunted with being a boring "mooch", just as he lashes out at

Freddie who has accused his imitation of Dickie of failing, his chosen objects merely "bourgeois". The brutishness of Minghella's Dickie explains Tom's violence as reactive, opportunistic rather than native: rich kids lash out when they feel uncomfortable, poor ones do damage rather to themselves. As Jude Law notes in his commentary, the theme of class is central for Minghella: "I adapted the novel from the point of view of class warfare" (AM commentary) and the voice of class assurance is given to his invention Meredith Logue who confides in 'Dickie' (actually Tom) her discomfiture with people not born to wealth. This both excludes and includes Tom, dividing him internally between the man he 'is being' and a real inner self whom he is – I shall return to this. It is Meredith too who offers him the first inspiration to act the part of Dickie, when he meets her on the way to Mongibello and, seemingly instinctively, tries on the persona of the other man, better suited to this female interlocutor. Every time she returns it is to reactivate the impersonation as reality.

Another key addition is the development of the figure of Herbert Greenleaf, Dickie's father, whose poignant impatience with his son's difference – "people say that you can't choose your parents, but you know you can't choose your children either" – explains his willingness to accept a violent and suicidal Dickie as the solution to the problem. Effectively he fulfils Tom's fantasy of being the son he should have wanted, denying the truth recognized by Marge by sewing up the male conspiracy (she is merely "his sweetheart" whereas Tom is "another fellow") and allowing Tom at last to unite in his single body, like a replacement child, "the brother he never had and the brother Dickie never had". And the last expansive adaptation is the development of the figure of Peter Smith-Kingsley. In the novel he is an "upright, unsuspecting, naïve, generous good fellow" (Highsmith 236) with "an Afghan, a piano and a well-equipped bar, [who] never wanted anybody to leave" (196), whom Tom knows he could never "be" because he does not look enough like him – and because the thought of impersonating someone so kind makes him feel "bitterly ashamed" (236). In the film he is the lover Tom Ripley should have had all along: musical in Tom's way, trusting and embracing, accepting Ripley at his best, and whom Tom murders in a paroxysm of tears knowing he has nowhere to go now but back into the dark.

By the use of these extensions, Minghella opens up the path of desire to mediation and complication. Another way that it is amplified is by the indirection of its aim. In *Gattaca* there is only one subject of desire; in *The Talented Mr Ripley*, desire has only one object, Dickie; and the others vie in creating ways of coming closer to him. Of all of them, Tom alone goes to the ultimate, cannibalistic length of taking on the object as his skin and disguise. It is surely the success of his cannibalism that makes Meredith and Peter recognize something in him that is more than he could have been before.

If Tom does not exactly want Dickie himself, what does he want that only 'being Dickie' can allow? We have already seen that he wants what Dickie has – what he takes on are Dickie's clothes, name and things: his "stuff", as Minghella puts it, or as Jude Law notes in the cast commentary: "the trappings of his lifestyle: great boat, great shoes, great rings, and watch and saxophone. To have a life where everything you look at is just sumptuous". These objects have an air of fetishism not only because they are another mediation, representing the wealth that makes them possible, but also because they can – must, by the interloper – be appropriated caressively. They are surfaces of desire – fine leather, velvet, cashmere. When briefly, on a train, Tom lays his cheek on the surface of Dickie, it is his lapel he reaches for, not his skin (as, later, just before strangling him, he lays his cheek on Peter's sweatered back). Dickie's clothes *are* his skin, they contain his solar glow exactly as Princess Diana's fabled outfits contained hers. The metaphor of sunshine, used both in the screenplay and in the director's commentary, is not a positive one: Dickie as played by Jude Law is "so sunny, such a sun-god", but unlike Diana's his skin is a reflective rather than radiant surface: his surface has a "metallic, cruel" quality, just as he is a semi-clothed "satyr" (crop-trousered man-goat, careless of others' desires because they offer themselves so unfailingly). By contrast, Minghella refers to Tom having to return to a dull texture, "the corduroy jacket" (AM commentary) when he gives up the Dickie persona to hide from the police (curiously, in Highsmith's novel, it is Dickie who wears the corduroy jacket, a reminder of how anachronistic glamour can be).

So: does Tom want what Dickie has, or what he wants? Or does he, rather, desire the way that Dickie wants? Like a foetus, Dickie scarcely knows desire, so quickly are his wants met by the world he glides through. Perhaps Tom wants to stop desiring. In Dickie he might be able to take a break from it; Dickie does not need to long for anything, it is at once his. He magnetizes the wishes of others, and this is what radiates from him, what he laughs off. Tellingly, when this momentarily fails, at the point of Silvana's suicide, he strikes out because he has no strategy for frustration. Tom, let us say, wants Dickie's ability not to waste time on desire, but to exist instead on the gleaming surface of things, himself *being* the surface of the things others look at. Unlike in *Plein Soleil*, this is not the triangulation of jealousy or envy (see Girard; also Sedgwick 1985 and Segal 1992: 59) but a wish to be a still point that is the indifferent site of other people's passions.

Can the manner (precisely not the manner born) be acquired? If it can, it is not by purchasing things – Freddie is right to disparage this as clumsy, "bourgeois" – nor even quite by what I have called cannibalism, because actually Tom does not ingest Dickie, he puts him on. Not like the gross 'Buffalo Bill' of *The Silence of the Lambs* nor the protagonist of Anzieu's 'L'épiderme nomade', but by acting out, literally, the gesturality of Dickie. As Lorenzaccio laments, you cannot act the externals without the internal impulses that make them appear. "The main thing about impersonation, Tom thought, was to maintain the mood and temperament of the person one was impersonating, and to assume the facial expressions that went with them. The rest fell into place" (Highsmith 114). Which is, finally, as much as to say that Tom already *is* Dickie because he can imagine him so accurately – this must be what is meant by 'talent'. He barely needs to imagine what it is to be as unimaginative as Dickie, already he can play it. But he can only 'be' Dickie when the latter is dead. And similarly he cannot play Peter because the latter is there, offering to supplement Tom as his accepting other. Tom Ripley does not want – or, he thinks, he must not want – any other. And yet, unlike Vincent, he has no choice but to remain mediate. I think this is what the ending is meant to convey: if his inner space exists at all, it is as a place where 'self' can no longer be shifted out into a gestural, careless surface, but is established as an inescapably inward, 'locked', secret, nasty thing.

What these two fictions suggest is that there is no way back from the assumption of another person's skin. The desire to be what Eugene painfully and Dickie effortlessly are is not only a homicidal one – the replacement child must be born at the cost of another child's death – but also suicidal: the other's identity is not a borrowed garment but a 'common skin' that tears you when its first owner dies. As Anzieu notes: "sub-tending both secondary narcissism and secondary masochism, we find the phantasy of a skin surface common to mother and child, a surface dominated in the latter case by the direct exchange of excitations, in the former by the exchange of significations" (AMP 65). In *Gattaca* signification predominates, in *Ripley* excitation; but in both, death is the outcome.

To return again to the extra-diegetic question: why Jude Law? – and why do we (if we do) concede to the unlikely fiction of the gawky, angular Matt Damon, even if he's briefly somewhat less gawky under Meredith's flattering gaze, being taken for the other man? Yes, he brushes his hair more smoothly, plays the rich boy just as fully as he appreciates opera and antiquities – in these things *plus royaliste que le roi*. But he never glows with the pollen-like sheen or insolent glances of his object – or of other glamorous figures like Lafcadio or Peter Pan. Is it that people mistake who Dickie was (looking to him, too, to have a body that clothes a 'self') in a rather similar way to how they mistake who Eugene is: an occasion, in both cases, for desire – and if it is, then perhaps this, to answer provisionally a question I posed several pages ago, is the use of beauty.

Being John Malkovich differs from both my other films in being not a modernist myth of masculine identity but a postmodernist narrative of multiplying desire.[22]. Its predominant metaphors for 'being' someone else are puppetry and acting, and it follows two men through their experience of exchanging these roles while, intersecting with this, another myth about how to desire brings together the two female protagonists to undermine an original heterosexual triangulation. Thus the version in play here is that of quadration.

Four-way structures in fiction have hitherto generally been a superimposition of two traditional triangles. Consider that of Goethe's *Die*

Wahlverwandtschaften [*Elective Affinities*] (1809). Here a central couple is perturbed by the entry of two desiring strangers – rather than one, as in the usual novel of adultery. These strangers both, in different styles, invade the couple in oedipal mode, prising open a stable attachment. In Henry James's *The Golden Bowl* (1904), the base pair is a father-daughter bond invaded by another couple whose liaison is secret and whose adulterous motive is disciplined by the daughter's refusal. In neither case is there a transaction among four elements but a variation on the theme of adultery, played on a more complicated chess-board with inter-generational nuances. But in *Being John Malkovich* we see the idea of multiplication running comically wild and suggesting the possibility of exponential futures. What we will find in the four protagonists is, respectively, the masculine principle embodied in Craig, twists of gender finalized in their original form in Lotte and John,[23] and, in Maxine, the feminine principle with which, Russian-doll like, the film ends.

In taking the four protagonists in turn, I am looking for the ways in which the base plot, that of Craig, runs towards its nth point. Craig [John Cusack] starts us off in the conventional place of the artistic adulterer. What can he do from inside John Malkovich that he cannot do when he is not? A puppeteer in the Kleistian mode, who manipulates his dolls "with incredible subtlety" (Kaufman 2000: 3), he shows a pair of figures enacting the passion of Heloïse and Abelard with such sexual explicitness that a father beats him up in the street. His libidinal engagement with the puppet figures is so direct that we see it as playing out all the longings his sweet dull marriage does not fulfil – in proper artistic sublimatory mode ("you're making him weep but you yourself are not weeping!", he/Malkovich accuses a pupil in a master-class shown later in a TV documentary) he uses fiction to channel his desires. Less complex than Flaubert's, we assume, his creative manipulation both diverts and preserves fantasy: the subtler the play of strings, the more complete the control, the more an author takes away a segment of him-/herself and exerts power over it – takes away, because this segment is excepted from the negotiations of everyday life. This will continue to be Craig's dilemma.

The day he joins LesterCorp as a lightning-fingered filing-clerk, he is captivated by a colleague, guesses her name, is rebuffed on their brief date and goes home to make a Maxine puppet who will murmur to the Craig puppet (unlike her original):

'Maxine'	Tell me, Craig, why do you like puppeteering?
'Craig'	Well Maxine, I'm not sure exactly. Perhaps the idea of becoming someone else for a little while: being inside another skin, thinking differently, moving differently, feeling differently.
'Maxine'	Interesting, Craig.

Craig's motive for entering the portal, then, will primarily be the wish to "become someone else for a little while" in the way that puppeteering has almost allowed him, because what he desires is Maxine [Catherine Keener]. To 'have' Maxine he is willing to sacrifice everybody: his wife, John, his own separate existence, and even possibly Maxine herself when she is kidnapped by Dr Lester. It is not easy to work out whether his art, and the power that belongs to it, is only a means to this end, but that is implied by the fact that when Maxine suggests commercially exploiting the portal and later his skill at manipulating the Malkovich body, he abandons his earlier insistence on the purity of his art and follows this speedier means to gratification.

Once Craig has chased his rival Lotte [Cameron Diaz] out of Malkovich and laid claim to a surprised Maxine, he learns to puppeteer the body, first to "feel what Malkovich feels rather than just see what he sees", then to "control his arms, his legs, his pelvis, and make them do my bidding" and finally, in a nice parody of sexual zen, to "hold on as long as I want. Oddly, it's all in the wrists".[24] He ends still in love, still frustrated, "absorbed" inside the head of little Emily and trapped for ever, because, as Dr Lester explains to Lotte, those who enter a "larval vessel" are "held prisoner by the host's brain, unable to control anything, forever doomed to watch the world through someone's eyes".

The punishment of libidinous Craig, who betrays all those who love him and (worse?) his considerable artistic gifts, is to dwell eternally, disempowered, inside the body of the daughter of his two womenfolk, gazing at a lesbian primal scene. The comical nature of this radical exclusion is that it actually has him enclosed at the very centre of a set of Russian-doll containers, and it asks a few questions about the simplest or primary form of desire, a man's demand to 'possess' a woman. In his final state he fulfils the ultimate version of his wish: the phallic trapper entrapped, inserted into a feminine space he cannot manipulate. The only puppetry (glove puppets are, strangely, never mentioned in this fiction) is not to be had by nimble-fingered manipulation of strings from above, but rather by disappearing inside the puppet-object; Craig tries this when the opportunity offers, for his puppets were, after all, always the instruments of a phallic aim; but when he does, neither Maxine nor even Malkovich answers to his powers. He should have stayed outside – but that too would not have brought him what he desired.

Lotte starts out as the sweet frumpy wife, doling out nurturance to a family of assorted animals her husband cannot reliably name; by contrast to sassy Maxine, she is an 'indoor woman', contained in her chaotic home, waiting disconsolately for the wandering Craig. But then she goes inside Malkovich, and thereafter, in a series of comic declamations – "Don't stand in the way of my actualization as a man, Craig", or like GI Jane "Suck my dick!" – becomes seduced by the experience of living in a male body. One of Dr Lester's elderly ladies expresses it for her: "I've always wanted to know what it'd be like to have a penis. Now I'll know!" Her first visit inside finds Malkovich soaping himself in the shower; she utters little cries of excitement. In her next, he is reading from *The Cherry Orchard* into a dictaphone, and she murmurs "I want that voice", forces him by thought control (her own puppetry) to go to a certain restaurant and, once there, watching Maxine through his eyes, marvels: "I've never been looked at like this by a woman before". Unlike Craig, Lotte will, after passing through a genuine wish to be *also* the other sex, be loved as herself. Maxine sets up the challenge:

> Craig, I just don't find you attractive. And Lotte, I'm smitten with you, but only when you're in Malkovich. When I looked into his eyes last night, I could feel you peering out. Behind the stubble and the too-prominent brow and the male pattern baldness, I sensed your feminine longing, and it just slew me.

Lotte is thus fulfilled by being the subject whose subjectivity is the other's double-gendered object. When she returns into Malkovich to make love to Maxine, the script reads:

> Maxine   Oh my sweet beautiful Lotte!
> John   Yes Maxine, yes.
> Lotte   Oh Maxine, this is so right!

– and they consummate a mediate union that gratifies them both. It may be on this occasion that Lotte/Malkovich impregnates Maxine. Interestingly, John Malkovich the actor is never required to play Lotte; when she is in him, the two characters speak separately in their own voices, and they remain separate though simultaneous in the eyes of Maxine, as we shall shortly see. Lotte has her reasons for saying to Craig "We love her [...], John and I": firstly because it arrogates both the bodies he too wants to 'possess', but secondly because it is as the two of them simultaneously that she wants to desire – still herself, but herself "actualized as a man". After all what originally fascinated her about the Malkovich body was its having, in the form of the portal (shades of Jude Law in *Existenz*!) both "a penis *and* a vagina".

The trajectory that brings Lotte from Cinderella to Princess Charming has to go via a journey through an other-sexed body that gives her enough pleasure for her not to need to stay there; she 'has' and then loses Maxine, survives loss and anger and comes through with both genders triumphantly intact. That she ends up as the father not the mother of Emily is a proper riposte to her housewifely beginnings, but it is also because the two versions of femininity in this fiction can only find their happy ending by swapping styles.

In the figure of John we find the most passive of the four protagonists – precisely because he represents the alpha-male body that everyone thinks can attain their desires for them. Dr Lester has been watching and grooming him, it seems, since infancy; queues of sad fat people want to occupy his head for their fifteen minutes; Lotte has a "very unhealthy obsession" with him, and even Craig sees him as his "livelihood". The real-life actor John Malkovich lent his body and a camped-up version of his public persona, male-pattern baldness and all, to this combination of narcissism and humiliation.

It is, indeed, difficult to distinguish absolutely between the fictional John and the real-life John Malkovich, because the meaning of both as we view them in the film is that they are actors, both possessed and possessors of the characters embodied in their skin. He gets to play not only a bit of Shakespeare, a bit of Chekhov, John Cusack, Orson Bean, a puppet doing ballet, a host of simultaneous cameos, and himself – he also acts, dramatically, what it might be like to battle against one or many occupying personalities. In other words, John Malkovich has the fun of playing an actor who becomes a puppet. And the fun is significant, because it means that we cannot ask the questions we asked, earlier, about Jude Law and beauty. As an actor-persona, John Malkovich carries connotations of intellect and European sophistication (he has played Valmont and Charlus as well as Gilbert Osmond and, more recently, Tom Ripley), but no one would say he was the most beautiful of male leads; what is at stake here seems much more to do with masculinity as status. So what do we see John experience when he repeatedly finds himself reduced to a conduit, a vessel for other people's gratification?

His friend Charlie Sheen (played of course by himself) loves the idea of a Maxine as a "hot lesbian witch" whose fancy it is to call her sexual partner Lotte. But John himself is freaked by the feeling that "someone was just moving all the way through me, moving my arms, moving my hands, talking for me" etc. This is, then, a version of the pregnant man motif featured in a number of 1990s movies. But, let's not forget, it is usually the Schwarzenegger-type, the excessively masculine frame that is thus comically invaded. Observe, then, that the ability to ironize your masculinity is a sure-fire way of being assured of it. It once more separates outer husk/hunk from inner other. Like the drag artist whose version of the money shot only reassures, the 'possessed man' goes beyond humiliation. John is the inverse of Craig, finally removed from the whole arrangement; but real-life John Malkovich, full of his flirtation with the feminine, ends triumphant as the envelope of envelopes, his eponym and face dominating the movie for ever after.

An actor is, like a princess, the object of everyone's gaze – but is he desired? The answer, for both the John Malkovich and the two Jude Law figures, seems to be no: he is desired with, through or by; he is a puppet. It is interesting to wonder why Spike Jonze chose to make an apparently minor change to Kaufman's screenplay: he changed Craig's artistic self-reinvention from a Malkovich puppet to a Malkovich puppeteer. The logic of the original version is that actors are the instruments of other people's fantasy. The logic of the new one is more flattering: actors are themselves manipulators; this is, we have seen, John Malkovich's film.

In the intra-diegetic terms of the functions of desire, however, it is Maxine's. Maxine starts out as a cross, glossy model type as ready with a put-down as a come-on. She bedazzles both Craig and Lotte, not only as the object of their lust but as possessing, seemingly without effort, the smooth surfaces their chaotic bodies lack. She wears close-fitting monochrome outfits, sits with her legs comfortably spread, smokes, leafing magazines, mobile phone in hand; she is always who and where she wants to be. Her ideology is enviably simple and it seems to permit not only her but others too to satisfy desire:

> Maxine   The way I see it, the world is divided into those who go after what they want and those who don't. The passionate ones, the ones who go after what they want, may not get what they want, but they remain vital, in touch with themselves, and when they lie on their deathbeds, they have few regrets. The ones who don't go after what they want... well, who gives a shit about them anyway?
>
> *Maxine laughs. There is another silence. Suddenly, at the same moment, both Craig and Lotte lunge for Maxine and start kissing her passionately about the face and neck. They stop just as suddenly and look at each other.* (Kaufman 51-52)

What is universally desirable about Maxine – it attracts John as well – is this insouciant combination of natural selection and *carpe diem*. This suggests that in a more metaphorical way (as hinted by Craig's Maxine-puppet dialogue) the object of desire is another person we wish to be. But the metaphor goes a different way here: one does not put Maxine on but discovers her principle as one's inner unlived self. Maxine gives Craig the chance, after all, to abandon high-minded poverty for a bit of John Malkovich-like stardom and she gives Lotte the chance to impregnate another woman without changing sex. What does she herself want?

While Craig and Lotte learn new ways to act on desire, John and Maxine are possessed by the desires of others. It destroys John but it is the making of Maxine. Narcissistically she gains in proportion as he loses: he is the occasion of her being able to say ecstatically: "Have you ever had two people look at you, with complete lust and devotion, through the same pair of eyes?". She becomes the princess they all long for, but here and now, with the focus of the bedroom not the diffusion of the balcony: she is neither gazed up at nor down on, but looked straight at, the object of an endless POV. Just as the tunnel of the portal flows down into the telescopic view by which the screen represents what people see when they are being John Malkovich, so the sight of Maxine made available to Lotte or Craig is almost precisely the phallus-eye view into the beloved. Maxine is the person we may enter if we find the way how, who will always welcome us and never wish to be us. This is what excites and gratifies her.

Maxine exemplifies the magnetic energy of the woman who seeks a passive goal with a large "amount of activity" (Freud 1973: 149). She alone never goes into John Malkovich, nor seems to want to. Instead she is happy running the business empire he becomes with Craig inside him – "Craig can control Malkovich and I can control Craig"; for after all, "it isn't just playing with dolls [...] it's so much more, it's playing with people!" – but she is not simply the manipulator of manipulators, for in the end she is undone by love. This, I suppose, is because her function of focusing other people's desires into her own narcissism finally leaves her prey to a directionless gratification. Only when she finds herself pregnant is she, unexpectedly, filled with something. In Maxine we see the undoing of the outdoor woman who has given herself up too eagerly to passivity. While Craig/Malkovich lounges in front of his TV, she croons to her belly and strokes the Lotte puppet he hung up long ago beside his worktable. Pregnancy, this suggests, is the last and most satisfying way of 'being possessed'.

What becomes of the 'thing inside' after it is born is answered in a negative mode. The delightful healthy Emily who swims underwater to the closing credits, is a new version of the Malkovich phenomenon, containing now Craig and in the future probably Lester/Malkovich, Sheen, their wives and who knows what others. The outcome of multiple desire, embedded in its women's space, her body is in no sense a resolution. Why is this?

I would see the impossibility of the process as its main significance: in none of my three films does 'a third being' get definitively made. In that sense the housewifely metaphor holds for all of them: making, cleaning, desiring are all provisional. It is true that a reproductive dialectic informs all these fictions. When two, three or many come together in desire there is, it seems, always an implicit outcome to the act of 'merging'. In *Gattaca* and *The Talented Mr Ripley*, it is achievement – and death. But in *Being John Malkovich* it is neither.

The containing body is intrinsically uncanny to the observer because of the ambivalence of its declaration: at once autonomous and 'possessed', and possessed invisibly (see Tyler 2001). Inside seven-year-old Emily, reliably virginal, permitted by her two loving mothers to go amniotic in the safe space of a sunlit swimming-pool, we are shocked to discover sad Craig skulking unrequitedly, repeating "Maxine. Maxine. I love you Maxine". But I'd like to finish this chapter on another set of questions.

When Maxine tells Lotte she is the father of the child, she means not that Lotte's sperm impregnated her egg (that must have been Malkovich's) but that the conception – in a very ancient fantasy – was aroused by the desire of Lotte ('of' in both senses). If conception is traditionally 'an idea that a man has inside a woman, then Emily is Lotte's idea – Lotte as fantasising being a man but actually gravitating towards an all-female genealogy. But suppose that Emily was conceived when Maxine *thought* she was responding to Lotte's desire, but actually it was Craig inside Malkovich. In this case, we arrive at three questions: How many fathers and mothers does this child have? Is it the actual or only the fantasized desire of the other that makes us quicken? And is Emily daughter or mother to the puppeteer she carries inside her?

Chapter 8: Love

Bodies itch and glow; desire orbits and is magnetized. This chapter looks a little further at the spatial modalities of desire as it moves towards the body of the other. In the last chapter we saw one version of the desire to penetrate; in this we shall examine three further 'formal signifiers' of desire and how they are or are not consensual.

An object lesson: what is it we imagine getting to when we wish to get inside a beloved person? The protagonist of Sartre's story 'Intimité' [Intimacy] complains about the incompleteness of her husband's love:

> He loves me, but he doesn't love my guts, if you showed him my appendix in a jar, he wouldn't even recognize it, he's always groping me but if you put the jar right in his hands he wouldn't feel anything inside himself, he wouldn't think "that's hers", you should love everything about a person, their oesophagus and their liver and their intestines (Sartre 1939: 107)

Is there a contradiction between wishing to get into the other and imagining what we would find there?

> Maybe people don't love those bits because they're not used to them, if they saw them the way they see our hands and arms maybe they'd love them; in that case, starfish must love each other better than we do, they stretch out on the beach when it's sunny and pull their stomach out to take the air, and everyone can see it. (108)

A similar idea about the 'insides', though in a more sadistic tone, underlies David Cronenberg's *Dead Ringers* (1988). It is, of course, possible by such techniques as X-ray, ultrasound or CAT scans – or, more impressively by the motion-picture use of endoscopy – to 'see inside' our own or other people's bodies (on the normal ignorance of the inside of one's own body, see Fisher, Leder, Jacques-Alain Miller). In 1996, artist Mona Hatoum made the video *Corps étranger* [*Foreign body*], which moves from a caressive journey across the surface of her skin to take the viewpoint of an endoscopic camera inserted, in turn, into her throat and cervix and revealing her oesophagus, intestines and other viscera. But, as Laura Marks points out: "The question of identification in this tape is perplexing [...] Hatoum can 'afford' to treat her body as an object; the effect of this work would be quite different if it were performed with any body but her own" (Marks 2000: 190). A comic version of the inter-body story can be found in the form of a promiscuous gift in Robbie Williams' music video *Rock DJ* (2000), where the tattooed and muscular star, singing on an island-stage encircled by skating or ogling models, fails to interest the girl [Lauren Gold] even after removing the last garment, so he takes his striptease to its logical conclusion by ripping off skin, guts and buttocks and finally, rocking still, duets with her in just his bones. An traditionally tragic one is the obsession of Musset, whose Lorenzaccio we have already seen lamenting the impossibility of separating mask from flesh, with reaching below the surface to expose inner corruption. In the opening scene of *La Confession d'un enfant du siècle* [*The Confession of a child of the century*] (1836), the protagonist discovers his adored mistress's infidelity by peeping under a table-cloth; disillusioned, he embarks on a period of debauchery and observes:

> The fatal idea that truth is nakedness was in my head now all the time. I said to myself: the social world calls its face-powder virtue, its rosary religion, its trailing cloak propriety. Honour and morality are its two chambermaids; in its wine it laps up the tears of the poor in spirit who believe in it; it walks with lowered eyes while the sun is high; goes to church, parties and meetings; and in the evening, it undoes its robe and reveals a naked bacchante with the feet of a goat.
>
> But talking like this just made me loathe myself; for I sensed that if the body is underneath the clothing, the skeleton is underneath the body. (Musset 1973: 111-12)

The inside or underside, the real nakedness of self or other, is nothing but more body, unknown but surely incapable of speaking a final truth. Nevertheless we imagine we wish to reach something beyond the surface of the beloved's body.

Is this the wish, as in *Being John Malkovich*, to nestle inside the skin of other, and if so, what for? In order to penetrate a third object, the one they will enable us to invade – and what then? When Craig meets Maxine, he conceives the wish to become "someone else for a little while: being inside another skin, thinking differently, moving differently, feeling differently" – getting inside John allows him to get into Maxine, who does not desire him in his own body; but then the "other skin" he is inside is not hers but John's. Does he ever get inside Maxine? Does John? Does Lotte, delighted with the whole package of having her idea via John inside Maxine? Whose child is Emily?

More precisely, *who* is Emily? If her inside is Craig, then who is she? Oesophagus, liver, intestines – and Craig? But she is not a starfish, and what we see is only a little girl sunning herself and looking at her mothers, while out of her eyes there appears – to us, puzzled, ignorant and vouchsafed not the view but the sound of his 'other' voice speaking as she gazes – the sound of Craig's desire.

I will return in the final chapter to the afterlife of the beloved no longer beloved, to look at how we try to conserve the lost object by a variety of means inside or perhaps outside our self-without-the-other. In this chapter I want to translate something like the taboo on touching into three modes of desire, none of them properly penetrative and all of them ways (negative or positive) of *not* reaching that elusive and frustrating 'inside'. These three modes are zooming, hovering and the caress.

The first two, perhaps surprisingly, are parts of a single gesture, the motion-above that is flight. An example is Baudelaire's poem 'Élévation', in which, in a series of vivid images of movement, the poet imagines his "spirit" leaping up away from the earth and forging "avec une indicible et mâle volupté" [with an ineffable, virile delight] towards "les champs lumineux et sereins" [light serene fields]. But in the last two lines, motion is replaced by another spatial relation. Happy is he

> – Qui plane sur la vie et comprend sans effort
> Le langage des fleurs et des choses muettes !
> (Baudelaire 46)

> – who hovers over life and understands with ease
> the language of flowers and silent things!

Birds and other flying things are a central passion of Romantic poets: Hugo's verses are full of swans, doves, butterflies, eagles and other avatars of the poetic "songeur ailé" [winged dreamer] or his loved ones (Hugo 339). In Baudelaire they are part of a fascination with claustrophobia that focuses on the lowering skies and tide of roofs of 1850s Paris. And yet, as we see in all these poems, flying never reaches a goal. Vast skies are framed in the city by windows or balconies and swans paddle in dust; over the ocean, albatrosses soar only to be snared and mocked; even the last voyage of death cannot be imagined except as anti-climax: "La toile était levée et j'attendais encore" [the curtain had gone up, and I was still waiting] (Baudelaire, 'Le rêve d'un curieux': 122).

The excitement of the poem is, rather, in the repetition of take-off – what Leo Bersani calls "a kind of vertical leap of consciousness" (Bersani 24) – that is rehearsed in a cluster of prepositions or verbs of sudden movement. Zooming as a fantasy cannot be separated from the moment of departing from the ground. Birds take off by generating enough airflow to create lift or dropping on to an existing gust of wind. Aeroplanes build up speed by taxiing, again relying on headwind or high-lift devices to set up the first upward motion. Dumbo proves he is no ordinary elephant by becoming the staple of drunken imaginings. Freud identifies the dream or fantasy of flying as a typical phenomenon, especially in children:

> why do so many people dream of being able to fly? The answer that psychoanalysis gives is that to fly or be a bird is only a disguise for another wish, [...] a longing to be capable of sexual performance. [...] Whenever children feel in the course of their sexual researches that in the province which is so mysterious but nevertheless so important there is something

wonderful of which adults are capable but which they are forbidden to know of and do, they are filled with a violent wish to be able to do it, and they dream of it in the form of flying, or they prepare this disguise of their wish to be used in later flying dreams. Thus aviation, too, which in our days at last achieving its aim, has infantile erotic roots. (Freud [1910] 1985: 219-20; for a difference between this motif in the two sexes, see F-Int 516-18)

And Kafka's 'Wunsch, Indianer zu werden' [Wishing to be a Red Indian] (1913) traces in a single breathless if-only sentence a centaur-like zooming that loses spurs, reins, ground and, by the fifth line, even the horse. Something of the same fantasy surely underlies Anzieu's 1992 description of himself: "I have formed with my superego a couple united in the way a horseman is with his mount – and I don't know exactly which of us was the man and which the horse" (Parot & Richelle 257). As in Kafka, the imagined unity of two such different creatures out of their more complex interdependence as master and servant – elsewhere, we remember, he calls the horse, like free association, "man's most noble conquest" (ACon 7) – actually means that one of the two must disappear. There is here a defiant endorsement of the castration complex that I will return to.

In *Le Corps de l'œuvre* (1981), as we have seen, Anzieu identifies creativity as "the illusion of lightness" (ACO 12) and 'take-off' or 'lift-off' [*décollage*] as its essential first stage: this is what transforms creativity, a predisposition, into creation, an activity: "Most creative individuals are never creators; what makes the difference, as Proust says of Bergotte, is the take-off" (ACO 17-18).

The wish to zoom is, as 'Élévation' shows, not an aim towards an end. Once Anzieu gets on to the five stages of creation, he leaves *décollage* behind. But we have already seen how intensively he sites the possibility of creation in a model of the male body. Thus even if the 'anchoring' of word or code in the body or emotions is one of the feminine aspects of creation, as is the sense of "being penetrated by a strong idea or by a project she feels as firm inside her" (ACO 86), these exceptions only serve to confirm the essential masculinity of the creator. Indeed take-off in this theory is something akin to the moment when the foetus, female by default in its earliest stages, receives the hormone that makes it male:

> why does an individual, whom one knew to be gifted, whether he thought this of himself or not, suddenly or at the end of a long incubation, begin to write, paint, compose, find formulae, and in this way have an impact on readers, spectators, listeners or visitors? Why does he fly forth while others remain on the ground? (ACO 18)

The fantasy of flying is gratuitous, purposeless, either an act of sheer undirected joy or the premise for something else. (In this, we can contrast it with the weighted, awaited object of Rilke's 'Der Ball', which rises in order to fall.) To soar like Superman is a simple phallic image – but take-off is a rather more complicated one. As the metaphors from Baudelaire, Kafka and Anzieu suggest, the desire to fly forth is a wish to gain by losing. It is all about positive separation, but – as the terms show in both French and English – it is also a risk of ungluing or unscrewing, of removing, of being separated.[25] If what can fly is the phallus rather than the man, who is he when he is no longer anything but his desire to desire? The boyish bravado – "I'm youth, I'm joy [...] I'm a little bird that has broken out of the egg", cries Peter Pan when challenged by Hook – that dreams of sexuality in the form of flying is dealing with the fear of castration by a kind of preemption; but then what becomes of the self that feared?

It explains, I think, the Baudelairean insistence that "les vrais voyageurs sont ceux-là seuls qui partent / Pour partir" [the only true travellers are those who leave for the sake of leaving] ('Le voyage', 123): the fantasy of soaring or zooming is simply the fantasy of taking off without any next stage. Or rather, what it leads to is a corollary that is also almost directly its obverse. Let us now examine the second fantasy of sexual desire: that of hovering. If we return to the ending of 'Élévation' where the poet, once on high, uses his position to drift overhead understanding the language of silent things, we find that Baudelaire's term is 'planer', to hover or glide. Anzieu's term, borrowed from Proust, is 'survoler': to fly above. Both images describe a relationship of stable superiority, a God's eye view, conferring knowledge rather than pleasure, an ability that Baudelaire suggests is something like hearing the unvoiced speech of the inanimate (flowers as *bijoux indiscrets* born to blush unseen?) but which Victor Hugo and others would present as reading the world as book – even though as writers they have created the thing they read.

As fantasies, authorship and hovering are closely allied, then. They both confer a divine privilege – but over something that is only *fantasized* to have preexisted the leap. In a letter of 1852, after all, Flaubert defines the presence of the author in the text as being "like God in the universe: everywhere present and nowhere visible" (Flaubert 16). Of course, our image of what it might be like to be God is drastically conditioned by our position 'from below', and it is this tyranny of invisibility that the aspiring author longs to assume. We have already seen this fantasy played with in a quite cruel way by the mourners of Princess Diana: did we want her dazzlingly alive or properly dead? The author-fantasy is a wish to be immortal vis-à-vis a toyshop of mortal objects we can scorn and ironize – characters, readers, pottering about far below.

In fact, of course, the ones who actually are immortal (since they have never lived) are the characters: Flaubert's compulsion to ironize stupid Emma or Charles is surely an expedient based on envy. These infants of his wishful mastery are actually the easiest things in the world to master – impossible not to master. But they are also attempts at mastering readership (Emma embodies this, since she lives and dies by reading), and readers are much harder to control. The wish to be immortal, which the children of our imagination do not even have to form, so inconceivable is it for them to die, is something that only flesh-and-blood people can have, and they have it by seeking virtual readers who will, like corpse-brides, agree to make them virtual writers. Nothing could, perhaps, seem further from the body that makes it possible to have desires at all. But that would be misleading.

Like Anzieu, Sartre uses the term 'survol' in describing how Flaubert in fantasy rises up above the rest of the human race who have made him feel abjectly despised: after climbing in fantasy to the top of a high tower from which giant-like position he can despise everyone, there is a sort of rush of motion and "whether he has been snatched up from the earth or the futile planet has dropped by itself into the abuses of space-time infinity, the fact is that he finds himself *in the air*" (Sartre 1971: 1185). Or again, "all of a sudden, panting and sacred, he rises up above his torturers, above Nero himself: how small they look, these instruments of his glory. He hovers and looks down, from the ether, at the rag he has left behind in their hands" (1177). The rag, like Marsyas's skin, is the bodily thing left after the fantasy has disembodied him. But we should not forget that it is the bodily thing that produces the fantasies.

Here is another, less human but also less agonized version of hovering. Leconte de Lisle (1818-1894), whose poems are suffused with a fulsome remembrance of Réunion, the Indian Ocean island where he spent his youth, writes of jungle scenes in which the apparent peace of sleep contains the coiled menace of animal violence: far-off lions or elephants slumber in the noonday heat, a tiger "falls asleep, its belly in the air, and dilates its claws" (LLB 175); and the jaguar dreams, a proper Freudian *avant la lettre*, that it is plunging "its streaming nails / Into the flesh of terrified, bellowing bulls" (185). His birds are nobler: his albatross, unlike Coleridge's or Baudelaire's (and contrast the vulnerable swans of Mallarmé or Rilke: some poets like their zoology classically uncomplicated) does not plunge to earth but "tranquil amidst the terror" of a violent storm on high, "approaches, passes and disappears majestically" (LLT 67). It is in 'Le Sommeil du condor' [The sleep of the condor], however, that the full fantasy of hovering – the coexistence of extreme power with extreme stillness – is clearest.

The condor is a member of the vulture family. It is supposed to have various peculiarities: to be able to go for long periods without feeding and to flush pink when emotional; but the aspect that has made most impact, and was noted by Darwin, is its ability to hover for long periods without apparently flapping its wings. Leconte de Lisle's poem begins, like Baudelaire's with vivid prepositions of flight, and then observes "Le vaste Oiseau, tout plein d'une morne indolence" (LLT 166) [the vast Bird, filled with gloomy indolence] gazing down upon the map-like panorama of America. As night rolls in like a tide from the east, it waits "comme un spectre, seul, au front du pic altier" [alone, like a ghost, atop the lofty peak], until at last the darkness covers it. Then,

> Il râle son plaisir, il agite sa plume,
> Il érige son cou musculeux et pelé,
> Il s'élève en fouettant l'âpre neige des Andes,
> Dans un cri rauque il monte où n'atteint pas le vent,
> Et, loin du globe noir, loin de l'astre vivant,
> Il dort dans l'air glacé, les ailes toutes grandes. (LLT 167)

> He groans out his pleasure, shakes his plumage,
> erects his muscular, hairless neck,
> and soars up, whipping the acrid snow of the Andes;
> with a hoarse cry, he rises to where the wind cannot reach
> and, far above the black globe, high above the living star,
> he sleeps in the icy air, his great wings outstretched.

This is, of course, a fantasy of phallic absoluteness: permanently tense, permanently relaxed – the ballet of male desire. But, as we have already observed, the ideal relies on failure: not simply on the logical impossibility of this fusion of extremes, but also on a different, psychic impossibility. In relation to Baudelaire's sudden switch from zooming to hovering, Leo Bersani observes:

> The emergence of an erotic esthetic will also involve the eroticizing of knowledge. But in early poems such as 'Élévation' and 'La Beauté', the sexual imagery is merely juxtaposed with the epistemological claims. In 'Élévation', the description of the poet's spirit plunging beyond the confines of the "starry spheres" suggests sexual penetration [...], but this erotic "rising up" seems to have no effect on the nature of the poet's comprehension of "the language of flowers and of silent things". An effortless serene understanding is unaffected by the erotic energy of the leap into understanding. (Bersani 25)

My own view is that these contraries are disconnected in a rather different way. The erotics of the flying fantasy is three-fold. If we trace it in reverse, the end-point of hovering stands both for the *survol* of superior knowledge, control from on high, and for the erectile tension that has become a sort of immortality or grace. Before this, the effort of desire is expressed in the fantasy of zooming, reaching-towards. Before even this, the initial movement is a taking-off, the initiative of excitement that lifts. Each one of these actions is, separately and together, a tracking-forth of the excitement of castration. Like 'escape velocity', the most extreme and deathly version, or Vincent's aimless aim of going into space, they are all fantasies of distance.

In Anzieu's citation from Proust, the relation of take-off to hovering that represents Bergotte's creativity is a sort of zigzag: "In order to travel in the air, it is not the most powerful automobile that is needed but one which is capable, by sheer ascensional force, of ceasing to run on the ground and cutting across the line of its horizontal speed with the vertical" (Proust [1918] 1954a: 554). Bergotte's talent may be nothing very special in itself, despised by family friends in Rolls Royces, but it has this capacity: "from inside his modest machine which had at last 'taken off', he hovered above them [*les survolait*]" (555; ACO 17). Carefully examined, the first motion is horizontal, the second vertical, the third again horizontal, but no longer moving forward, for the relation of superiority is not directional but static. It is all about separation. This knowledge is, *pace* Bersani, still erotic, but an erotics of distance, coolness born out of heat.

The whole point of this fantasy is its inability to touch. The fact that it must not come to an end means that it is, effectively, all end. It is castratory because, ultimately, the machine hovers alone isolated from its origin. The obvious corollary of the condor – that patient predator – is the modern bomber-plane. Its association with death may be suicidal, like that of Yeats's Irish airman in 1919, driven on high by "a lonely impulse of delight" (Yeats 184), very similar to that of Saint-Exupéry's heroes experiencing "the mysterious labour of a living flesh" (Saint-Exupéry 23); or it may be homicidal like that of Marinetti, who writes in *The Battle of Tripoli* (1912) of the pleasure of bombing without needing to dirty his hands. But ultimately it goes out beyond the flesh, representing the extreme 'clean' violence of the *survol*: brains without bodies (see Jay 90, 213 and 387). In 1921, with remarkable prescience, Marinetti wrote of the possibility

– like Kafka's Red Indian fantasy – of the violence of hovering imagined at the furthest extreme from bodily presence:

> Phantom-aeroplanes laden with bombs and without pilots, remote-controlled by a 'shepherd' aeroplane. Phantom-planes without pilots which will explode with their bombs, which can also be guided from the ground by an electric control-panel. We will have aerial torpedoes. One day we will have electric war. (Marinetti [1921] 1985: 121)

Anticipating the tactics of today's aerial bombardment, and the very reverse of our contemporary suicide bombers, these masculine fantasies of desire are both self-separation and separation from the other. Consummation, it seems, is neither sought nor achieved; but there is no loss either, because the 'other' – land viewed from above, flowers and other mute things – is actually much too far away to be heard, seen or touched.

So what happens when fantasy *does* touch? There are essentially two ways of thinking about the skin as the site of erotic experience. The first is the solipsistic one in which the subject's skin (or whole body) is a focus of sensations that are in themselves intense or voluptuous but do not imply an encounter with another subjectivity. Freud's theory of psychosexual development exemplifies this position insofar as the infant experiences and seeks intensities or oceanic calm which seem to originate in and return to its own body; the other is not intuited or, in a sense, necessary. Lacan's pessimistic view of desire and eroticism follows the same essential line, as do the sensation-based skin-bodies of Deleuze and Guattari (1980), Lyotard or Lingis (1983 and 1994). This monadic version of sensual experience is an essentially autistic worldview, and the fantasies of zooming and hovering are part of it. At the other side of psychoanalytic theory are those thinkers for whom, as for Winnicott, "there is no such thing as a baby": only the mother-child couple (Winnicott [1952] 1984: 99), for whom nothing precedes the social. This base unit of two makes sensation intrinsically an encounter – either an encounter of two autisms implicitly destined for conflict or an encounter of two subject-object complexes capable (on a good day) of creating something potentially 'consensual'.

I am going to look now at the erotic gesture most clearly connected to touch: the caress. I want to examine the fantasy of the successful encounter through, with and perhaps beyond the body's surface that we call sexual love. Desire is a component, but not the only element of love: the wish to be held, in Anzieu's sense, is central. I shall look at five theorists, first in chronological order, and then grouping their observations together to support a final line of argument: Sartre, Merleau-Ponty, Lévinas, Irigaray - and Anzieu.

In *l'Être et le néant* (1943), Sartre has much to say about bodily fantasies of inside and outside - his long discursus on the viscous is a model of the horror of a feminized state between solid and fluid. Discussing the language we use to think about knowing [*connaissance*], his Acteon complex moves beyond the wish to know by hunting or "penetration" towards that of eating, and his Jonah complex is the fantasy of existing "indigestible digested" (SEN 625) inside the body of the other. Like the desire to know, the desire to appropriate is always frustrated. One instance is sport - specifically skiing, which illustrates the wish to possess a surface that remains opaque and impenetrable - "by my activity as a skier, I modify the matter and meaning" (628) of the snow, by a "sliding" [*glissement*] (629) that is "the opposite of taking root". Like art, science and play more generally, the gratuity of sport makes it appropriative in this "sliding" mode.

Desire too seeks to appropriate, but it too cannot go beyond the smoothness of a surface that "re-forms itself under the caress like water after a stone has passed through it" (625). More significantly, desire is always that of a body for another body, through which both discover themselves as flesh:

> Everyone is disappointed by that famous saying: "[love is] the contact of two epidermises". Love is not meant to be mere contact; it seems that only man can reduce it to a contact, and when that happens it loses its true meaning. The caress is not a simple floating touch [*effleurement*]: it is a

> *fashioning*. When I caress another person, I create [*fais naître*] their flesh by my caress, with my fingers. The caress is that set of rituals that *incarnates* the other. [...] The caress creates the other as flesh both for me and for themselves. [...It] reveals the flesh by divesting the body of its action, splitting it off from the possibilities that surround it (430)
> [...] In the caress what caresses the other is not my body as a synthetic form in action, but my fleshly body which creates the flesh of the other. By means of pleasure, the caress is able to create the body of the other both for them and for myself as a *touched* passivity, in the sense that my body becomes flesh in order to touch their body with its own passivity – in caressing itself against it rather than caressing it. This is why the gestures of lovemaking have a languor that one might almost call studied: it is not so much that we *take hold of* [*prendre*] a part of the other's body but that we *bring* our own body up against the body of the other. Not so much pushing or touching, in an active sense, but *placing up against* [*poser contre*]. [...] By *realizing* the other's incarnation, the caress uncovers my own incarnation to me. [...] I make the other person taste my flesh through their own flesh in order to make them feel themselves being flesh. In this way *possession* is revealed as a *double reciprocal incarnation*. (431)

Sexual desire, here, brings the body to the fore in a way that its everyday existence, a means of enacting projects in the world, cannot. It creates two selves of flesh. It does this because of the peculiar 'impenetrability' of the other's body, the caress being closer than an *effleurement* but further off than a penetration: a placing-up-against that slides briefly along the smoothness of the other's otherness. Unlike the appropriativeness of knowledge, sport or art, however, desire creates, through the caress that makes them flesh, an encounter of two bodied freedoms.

Sartre is famous for the pessimism of his theory of inter-human relations: I will not repeat here his arguments of mutual destructiveness, shame and masochism. What I am interested in is the less often cited consequence of these two facts: in the face of the doomed wish to "assimilate the other's liberty to myself" (405), what are the consequences of the mutual creation of each other's flesh in desire and the necessary passivity of the love encounter?

"In love, [...] the lover" (407-08) – I have retained Sartre's masculine pronoun here in order to preserve the singular though what he says is not conditioned by gender

> wants to be everything in the world to the beloved. This means that he is putting himself on the side of the world; he resumes and symbolizes the world, he is a *this* that contains [*enveloppe*] all other *thises*, he is and agrees to be an *object*. But on the other hand, he wishes to be the object in which the other's freedom agrees to lose itself, the object in which the other finds their second facticity, their being and *raison d'être*, the limit-object of transcendence, the one towards which the other's transcendence transcends all other objects but which it cannot itself transcend. And everywhere, he desires the circle of the other's freedom; this means that at every moment, in the other's acceptance of this limit to their freedom, he wants that acceptance to be *already* there as its own motive. He wants to be chosen as an end that is an *already* chosen end. This shows us what the lover fundamentally wants from the beloved: he does not want to *act* upon the other's freedom but to exist *a priori* as the objective limit of that freedom (408)

What I want when I love, according to Sartre, is to be "the object by whose proxy the world exists for another; in another sense, I am the world. Instead of being a *this* standing out against a background of world, I am the object-background against which the world stands out" (409-10).

In *Phénoménologie de la perception* (1945), Merleau-Ponty includes a chapter on the sexual body, but essentially it follows either Sartre's or Freud's line without much development. But in the posthumously published *Le Visible et l'invisible* (1964), a key essay on 'Interlacing – the chiasm'

(MPV 170-201) describes the essential doubleness of the sense of touch: "at the same time as it is felt from the inside, my hand is also accessible from the outside, itself tangible, for example, to my other hand, [for] it takes its place among the things it touches and is one of them" (174). In the same way as touch is doubled, vision and the experience of the body itself are simultaneously the subject and object of sensing; this is the chiasm, an "reciprocal insertion and interlacing of one inside the other" (180). The body and the world it lives in are like a pair of mirrors, two surfaces that form a couple more real than either of them taken separately.

This intrinsic reversibility means that the reciprocation of a human couple touching each other is already preempted in the individual; Glen Mazis uses this to announce a sexual intimacy in which "the distinction between activity and passivity dissolves [and] the situation of desire gives the caress the opportunity to maximize and heighten the reciprocity of touch" (Mazis 148-49). But Merleau-Ponty is more guarded than that: in this essay he raises but does not carry forward the possibility of an "'intercorporeity'" (MPV 183), a "synergy" (185) or "transitivity from one body to another" (186) which would be what I call "consensual", arguing instead that we only extrapolate the other's experience as being as multiple as our own and embedded in the same kind of world.

There are other clues, however. In the notes of 1959 and 1960 appended to the essays of *Le Visible et l'invisible*, the idea of the chiasm reappears extended to an intercorporeal realm: "the chiasm in place of the For-the-Other: it means that there is not only self-other rivalry but also co-functioning. We function like a single body" (264). This is posed as a response – not so much a resolution as a transformation – to the "problem of other people" (316) and linked in an undeveloped way to the idea of "carnal relations [...] Entwining". (317). There is thus (in note form) the beginnings of a theory of consensuality via the caress, but no more.

If we return to the earlier essay, though, we find a striking paragraph inserted in the middle of a discussion of the "solipsistic illusion" (186) of the bodied subject:

> For the first time, the body no longer couples with the world, it intertwines with another body, applying itself carefully to it

> with its whole expanse, tirelessly sculpting with its hands the strange statue which, in its turn, gives everything it receives, cut off from the world and its aims, occupied with the sole fascination of floating in Being together with another life, making itself the outside of its inside and the inside of its outside. And then at once, movement, touch and vision, applied to the other and to themselves, head back to their source and, in the patient, silent work of desire, commence the paradox of expression. (187)

This scene assumes the psychic isolation from the world, the mutual exchange of sensations and inside/outside blurring of the couple contained in the space of physical love. It also echoes, in the 'careful' image of the statue, something of the obsessions discussed in chapter 6, aestheticizing Sartre's concept of the caress making the self and other into flesh. The key term is 's'appliquer', which appears both in reference to the person of the lover working on the "strange statue" of the other's life and also to the senses working among themselves. The ending of this passage seems just about to point forward to a development of the "paradox of expression" – but it never happens, the argument at that point turning elsewhere. What this paradox might be, though, is the *active* relation of consensuality that would supplement Sartre's passive one. Again, I will come back to this.

Sartre and Merleau-Ponty assume a male subject of desire but neither of their theories actually requires it. Lévinas, on the other hand, surely the least macho of male philosophers, illustrates how the theory of the caress can fall into the trap of gender. An extreme passivity characterizes the fundamental relation between the subject and the other that he calls "sensibility" (LAQ 30-32). It combines exposure, patience and responsibility, a kind of denudation beyond the skin: "uncovering oneself – that is, stripping oneself of one's skin – sensibility at the edge of one's skin, at the end of one's nerves [*à fleur de peau, à fleur de nerfs*]" (31). This sensitivity to the demand of the individual other is a recognition of his or her vulnerability, which resembles my own, but it is an equivalence that is not a commonality:

> The relationship with the other is not an idyllic and harmonious relationship of communion or a sympathy in which we put ourselves in the other's place, recognizing them as resembling us but external to us; the relation to the other is a relation with a Mystery. It is their exteriority, or rather their alterity [...], that constitutes their entire being. (LTA 63)

This is what he describes as "the face-to-face with the other" (67).

The erotic encounter, on the other hand, perhaps because it is not a question of faces and light but bodies and darkness, is a different kind of alterity, one that is "absolute":

> Is there a situation [in civilized life] where the alterity of the other appears in its pure form? [...] Might there not be a situation where alterity would be carried by a person in a positive sense, as their essence? What is the alterity that is not purely and simply part of the opposition of two species of the same genus? I think the absolutely contrary contrary, whose contrariety is in no way affected by the relationship that can be established between it and its correlative, the contrariety that permits this term to remain absolutely other, is the *feminine/female* [*le féminin*]. (77)

In this theory, the position of the implied 'I' is based in its difference from an absolute other gendered feminine. There is much debate over how far '*féminin*' must mean women and how far it could be a circular logic about the other's opposition to the implicitly masculine subject (many commentators, such as Lingis 1985, Davies or Moyaert, use gender-double pronouns; those arguing the reverse include Beauvoir 16n, Chanter, Irigaray in IEDS 173-99, 1990: 911-20, and IED 48-55; and Derrida 96). I prefer to slice this Gordian knot and agree with Irigaray that there is an absolute sexism here that cannot be by-passed. But I would link this with the undeniable femininity of the overall posture of Lévinas's thinking, whose interest rests in the Jewishness of its attitude to gender (see Katz 2001 and 2003). Like the Jewish God – not bodied, not familial through a divine-human adultery, like the Greco-Roman or Christian deities, but a potter who fashions humans out of red earth (see Meyers 81-3) in imitation of himself – these moves are all cooptations of the activities of the female body (reproduction, nurturance, sustenance) onto a non-bodied male.

In Lévinas's theory, the caress is the encounter between a human, bodied version of this male subject and his female object. Just as we saw both Sartre and Merleau-Ponty stepping out of character and sometimes even off the logical path to represent the night-side of their argument, here we find an unexpected tone – what distinguishes it from both Sartre's carnalizing and Merleau-Ponty's sculptural touch, is the uncharacteristic element of the chase. And yet this is not a pursuit exactly but something more tentative, like an appetite:

> [it is as if] the caress fed itself on its own hunger. The caress consists not of grasping anything, but of calling upon something that is constantly escaping its form towards a future – never future enough – upon something that runs away as though it were not yet. It searches, it delves. It is an intentionality wanting not to unveil but to seek: it walks towards the invisible. In a sense it expresses love but suffers from the inability to say it. It is hungry for that expression, in an ever-increasing hunger. [...] Satisfied, the desire that animates it is reborn, nourished in some fashion by what is not yet, bringing us back to the ever-inviolable virginity of the feminine. (LTI 288)

We left Merleau-Ponty on the edge of an argument about expression – the term returns here, in an unexpected amplification of the image of hunger. This is the meaning of Lévinas's Acteon complex: the caress is hungry for its own expression of a love it cannot say.

Lévinas's carnal hunger is indeed the inverse of his theory of responsibility towards the other. The female – animal-like, childlike, "violable and unviolable" (289), both pitiable and provocative – provides a pretext for him to go "beyond the face" (291, 296) into "exorbitant ultra-materiality" (286). As virgin, she "remains impossible to catch, dying without murder, swooning, withdrawing into her future, beyond anything anticipation might be promised" (289). It is essential that she be both a she and yet faceless, because seeking her is the lonely pursuit of "non-significance" (292).

This essentially autistic experience has little to do with consensuality. And, although Lévinas agrees with our other theorists that the couple is cut off from the world of others - "Sexual pleasure is fundamentally contrary to social relations. It excludes the third party" (297) - one feels that there is actually no couple at the centre of his theory. This surely explains the urgency of supplying the terms of "fecundity" to restore the social [masculine] structures, bringing the child that will unite, divide and preoccupy, to a dangerous "egoism à deux". (298). But even before this, the pair have existed in an empty nucleus closed very much like a prison: "*The mutual action of feeling and being felt* that sexual pleasure achieves fences in, encloses and seals the society of the couple" (297).

I want to say one more thing in relation to Lévinas's rather disturbing image of the caress. It is hard to see this as a representation, even of the most general kind, of two people making love. This is partly, as we have seen, because there seems to be no other in the picture, partly because there seems to be no body, and partly because 'the feminine' seems such a crude amalgam of all the clichés of nubile womanhood. But perhaps that is because the image is actually quite a strange combination of fantasy and timidity. For all the description of a couple as cut off from the world, the world of this "intentionality walking towards the invisible" is a space without borders, the wild place of desire and risk.

In Irigaray's early writing, 'the fluidity' of women's sexuality is in danger of being subsumed into men's sexual purposes: "man needs an instrument in order to touch himself: a hand, a woman, or some substitute" (IS 288). Caressive love between women overrides this kind of instrumentality, as touch becomes a 'through' rather than a 'by' and achieves consensuality: "You don't 'give' me anything when you touch yourself or touch me: retouching yourself through me" (ICS 206); "you touch me, all of me at the same time. In every sense" (209). But in her later work, following the violence of her portrayal of mother-daughter inseparability, only an equivalence between two different sexes will do. First, men's use of women as internalized containers (we get an echo of Lévinas's sealed prison-cell) has to be refused:

> Your body is my prison. But since you make me yours from the inside, as you draw on my sources from the inside of my skin, I am unable to wrap myself in it again to go back outside. (IPE 17)
>
> [...] And you meet me only in the space you have opened up for yourself. You only ever meet me as your creature – deep within the horizon of your world. Within the circle of your future. This shell-shelter that protects you from an outside-you that questions you about the material in which you have built yourself this house (57-58)

Women are flattered by men's dependency and offer a different kind of containment: "You filled me with your voids. You filled me up with your lacks. Fortified by being their cure, I brought you my most precious possession: my hollows. You became wide open, I was full" (74). But this is destructive for both parties: "love is the becoming that appropriates the other by consuming them, introjecting them into oneself until they disappear" (32). When it is right – and this is presented in many pages of ecstatic prose – it is the exact balance of two equal and different subjectivities.

Just as, in her riposte to Lévinas, she argues against his "autistic transcendence" (IEDS 193) by adding to the scene a female lover [*amante*] as active as the male one, so the maintained gap between the two lovers is essential. Irigaray rejects the temptation of "the dense cloud of 'us'" [*la nuée du nous*] (ICS 25), preferring to preserve a "veil" (23) or "clearing" (33) between them and the cultivation of perception rather than sensation.

Irigaray's representations of the caress tend to be programmatic, even prescriptive. It is described in a litany of epithets – "awakening", "gesture-word", "incantation", "praise", "invitation to rest", "gift of security" (50-53) etc. – all dependent on the maintenance of difference, "an act of intersubjectivity, a communication between two subjects" (54). This may well seem excessively calm, in comparison with her more "elemental" writings. But here is a passage from an earlier essay in which Irigaray is rereading Aristotle's *Physics* (a description of the laws of material 'bodies' in space) in relation to gender and the human body. It seems to me to represent very exactly the movement from the zooming-hovering fantasy to the process of caress:

> The *motion-towards* [*transport-vers*] and the *reduction of the gap* [*intervalle*] are the movements of desire (even by expansion-retraction). The greater the desire the more it tends to try to overcome the gap while at the same time maintaining it. A gap that might be occupied by the transformed body? What is at stake in desire, the cause of the motion, is the wish to overcome the gap. The gap is reduced almost to zero in the contact of skins. It goes beyond zero when the contact passes to the mucous. Or by transgression of the sense of touch through the skin – the problem of desire being how to remove the gap without removing the other. Since desire can devour place, either by regressing into the other in the intra-uterine mode, or by annihilating the existence of the other in various ways. For desire to survive, there must be a double place, a double envelope (IEDS 53; her italics refer to Aristotle's terms).[26]

What emerges from this tour around the philosophy of the body in love is a collection of tentative theories: in Sartre, the caress creates a psychical membrane for the lover who is held by becoming the world-as-object for the beloved; in Merleau-Ponty the lover creates and becomes a "strange statue" contained in a communicative space; in Lévinas the lover seeks a feminized other who can never be found; in Irigaray lovers suspend each other enveloped in an ideal balance through skin and mucous. How does all this bring us back to Anzieu?

Anzieu wrote little about the caress, and it is never theorized in detail in his work; indeed it has been argued (see Évelyne Séchaud in AMP 8, 10, 12) that sexuality plays a relatively small role in his theory of the *moi-peau*. Certainly the taboo on touching is far more important than its obverse (see

APA 145, AMP 161-73) and his caution in relation to touch-therapies (APP 66, 70-71 and 75) illustrates the importance of using psychoanalysis to replace, not reproduce, the satisfaction of being touched physically: "one can touch with the voice" (APP 75). But outside the consulting room, bodies do meet. The word caress is mentioned in two kinds of place. The first is in Anzieu's fictions or the fictions he cites, generally in an auto-erotic context, often associated with disability or sickness (for the first group, see ACon 141, 198 and 201; the second AMP 176, ACD 218, ACon 30-31; AEN 14, 16-18 and 22). The second is as the "massage [as] message" (ACD 203) communication of mother to baby which leads to a more general "eroticization of the skin" and thence to a "diffuse pleasure" (ACD 213) that precedes or replaces genital orgasm.

In relation to sexual practice, the key question is what is meant by "skin pleasures" (ACD 203), the state in which "the whole skin is potentially a vast erogenous zone" (APs 153). Like Bollas, cited here in chapter 5 (103), Anzieu identifies this as an essentially 'preliminary' practice, by which he seems to mean an incomplete because non-penetrative act: that of the narcissistic hysteric, that of heterosexual 'foreplay' or that of lesbians. On the other hand, it is the communicative stimulation of the mother's hand or skin on the child's skin – to which we shall return – that lays down the possibility of the skin-ego that makes sexual fulfilment possible: "genital sexuality, including auto-erotic, is accessible only to those who have acquired a minimum level of security in their own skin" (ACD 203) and "experienced the erogenous zones on that skin" (AMP 246). Annie Anzieu holds to the view that women's sexual experience is derived from "the mystery of internal tactility" (AA 28), a "pleasure of the orifice that [is] specific to femininity" (49). In this view, women's pleasure is tied to the internal, to "penetration" and "the attraction of the hollow to the object" (48); she keeps a strictly Freudian line in adding that "in women, sexual sensations are, without any possible doubt, more internal than external. From the evidence of clinical observation, one might even wonder whether what people call clitoral orgasm is not actually the external displacement of the possibilities of internal pleasure" (77). But even this is less clear-cut than it might seem because, as we have seen, the vagina "also functions towards the outside. A place of exit: things flow, are born, discharge or slide out of the body. Milk, period, baby" (51). Indeed, this outflow is crucial: unless closed by "the object of pleasure", a woman's orifice is subject to uncontrollable outflows and "the liquid creates the conduit".

Didier Anzieu takes this development further and turns it around towards the man's experience when he connects sexual practice to "the affective origin of the mental relationship of figure to ground" (ACD 215):

> This leads to a rethinking of the notion of sexual pleasure as such. The automatic image of coitus as the male organ penetrating 'inside' the female body is actually deeply imbued with the archaic fantasy of the baby merged inside the mother's womb. In reality the vagina is not an internal organ but a deep fold [*repli*] of the epidermis, an 'invagination' of the skin. The pleasure of the exposed glans and the soft skin of the penis rubbing against warm, moist mucous membrane of the vagina – which is the condition of the release of orgasm – constitutes a synthesis of two 'primal' pleasures: the skin-to-skin contact of the newborn with its mother and the insertion of the nipple or teat into the mucous membrane of the lips, tongue and palate, followed by a warm, moist flow of milk.

Four points in particular are interesting here. The first is that, even though Didier Anzieu is describing sex from the opposite angle to Annie, "internal tactility" remains a mystery. There is, in fact, no 'inside': as we saw in chapter 3, the vagina is like all other orifices in being merely an in-turned segment of the skin endowed with especial sensitivity – "an in-folded leaf, not a hole" (APP 97). In the earlier instance, this fact was associated with love's "paradoxical quality" (AMP 32, cited here in chapter 3, 73) of being at once psychically deep and epidermally close; in this later discussion, it is connected to infancy. The second point is that the culminating image of sucking and drinking combines in itself the two angles of container and contained, 'masculine' and 'feminine': it is the baby's body that provides the warm orifice, the mother's that releases the fluid which delights it. This is the reason, perhaps, for the third point; in combining these two sides of the sensual moment, Anzieu is also 'merging' two very different stages in the infant's experience: the neonate's experience of suckling with the foetus's experience of being in the womb. The sexual experience calqued on the fantasized recollection of an inside/outside relation with the mother's body is both result and basis of a particular version of the fantasy of the common skin. The fourth point is to bring the whole discussion back to its context – the figure-ground structure that underlies mental activity: "The orifices [...] become erogenous zones – if they do – by being figures or points of intense, rapid pleasure on a ground of the global sensuality of the skin" (ACD 215). The figure is sexuality, the ground is consensuality.

With this in mind, I want to move on now to Anzieu's theories of love. Again, these are not at the centre of the argument of any of his books, they appear in margins, at endings, sometimes in multiple comparisons. In *Beckett*, for instance, written at the end of Anzieu's life, we find the following analogy: "The psychoanalyst and the writer register and respect the umbilical mystery of the crossing of unconsciouses between two people, whether in friendship, love, writing, reading or destruction" (ABe 248). Such an "umbilical mystery" links the love-relationship implicitly with the common skin, to which I shall return presently. But let us note for the moment the parallel drawn between the contexts of everyday emotions (love, friendship, aggression), psychoanalysis and the aesthetic. The whole of the Beckett book is, of course, imbued with the parallels between the practice of writing/reading and the practice of analysing/being analysed, just as the bodies of the real author and implied reader are shadowed by those of 'Beckett' and 'Bion'; a few pages earlier, we find the argument that style, like the ego, is both a skin and a "passageway" (243) and that in certain "privileged moments [...] one Ego functions for two. For the mother and her child. For the patient and his/her psychoanalyst. For the writer and his/her reader" (243-44 and, for the latter, see also ACO 12).

As Anzieu presents them, two people in love share both surface (sense) and depth (psyche). On the one hand they may be brought together by a need to conjoin "at points where their psychical borders are uncertain, inadequate or failing" (AMP 113) they may invest in each other in a "compensatory" mode (AMP 116), believing in the fantasy of being "glued by love" (AEN 38). On the other they may, like the mother-child couple, be jointly part of a "homeostatic system [...] closer to a pair of twins than a complementary dyad" (AMP 81). This idea of being twinned or suspended together is particularly characteristic of the young couple in love, who see themselves as having "a common ideal ego [...] fused inside a single psychic skin" (ACD 248). In the following passage Anzieu moves through two versions of the common skin fantasy, ending with a different kind of membrane:

> I prefer to think of the two lovers as linked in imagination by the fantasy of a common skin, a duplication of the common skin between the mother and the child – touching, holding or embracing each other all the time, aren't they expressing the need to share a single piece of cutaneous surface between them? This interface, in the position of a diameter internal to their couple, is a surface for the inscription of signs, an organ of direct communication, without mediation, which supposedly lets each of them know immediately the thoughts, impressions and affects of the other. In addition, the couple in love shut themselves inside a bubble, trying to function in a closed system, *in camera*, cut off as far as they can be from the demands and intrusions of the outside world. That is a different kind of psychic envelope from the surface of inscription, an envelope of protection against stimuli that is total, global and, in my view, double so that it can be more resistant: the lovers make it out of the imaginary skins of their respective mothers (or primary carers) which they took away when they separated from them. (ACD 254)

This fantasy includes the wish, described in *Le Penser*, to be "an introjectable kernel which spreads out into an envelope. This is a first paradox: that the other should be at one and the same time the centre and the periphery, the shell and the kernel" (AP 67).

Anzieu evokes "the double belief that the partner is the object that counts above everything for me and that they also have the desire to be that primordial object for someone, who happens to be me – just as the mother wanted to be this for her baby who, in its turn, put her in the place of being its object" (ACD 254). This exchange of objecthoods within a closed spherical membrane is an idyll of mutual passivity, freely accepted. It is, as we shall see, the common skin as reciprocally inclusive "in which each of the lovers, holding the other in their arms, envelops the other while being enveloped by them" (AMP 85). This idyll cannot persist, of course, at the same pitch and the couples who survive – without or beyond the usual "domestic scenes" (ACD 251-61) – are those who have learned how "to live progressively with disillusion [...], to accept a sufficient differentiation [and] to create together some kind of familial psychic apparatus" (ACD 249).

There are two last words that Anzieu offers on the theme of love. One we have already seen: at the end of *Beckett* he leaves a 'Finale' that includes the phrase "To sustain love in the gap between abandonment to the other and abandonment of the other" (ABe 289). For love to survive disillusion and even loss, this principle is essential. The other, some years earlier, appears at the end of the interviews with Gilbert Tarrab.

> – *I'd like to quote what you said in the notes you wrote in preparing for this book. You said about yourself: "In his public life, he liked three things: making people think, making them laugh and making them dream. In his private life, one thing mattered to him above all: being loved".* Being loved, yes – but what about loving?
> – That phrase came to me spontaneously. I thought when I wrote it that you'd ask me this question, and it's quite a poser. But I decided not to alter the phrase my unconscious had dictated to me. It didn't actually surprise me that much: all through my childhood and adolescence I lived with the feeling of not being loved by my mother.
> – *I don't want to go on too much about this, but your phrase has many levels. I'm trying to understand. The reader will also want to understand, I'm sure.*
> – I wouldn't have spoken to you about the richness, power and complexity of the feeling of love, as I did at the end of our sixth interview, if I hadn't had personal experience of it, but I do believe that this question, when it is broached in public or for an audience, must be treated soberly and

discreetly. That is my first comment. The second is that love may be used well or badly. Love is not simply generosity: one can crush or stifle someone with the poisoned gifts one gives them. Love is not just accepting the other person's suffering – including the suffering the two people inflict on each other, which sometimes comes to fill the whole stage of their drama. Love shows that it is intelligent when it helps to create, for the child, the friend, the partner in life, a supple and firm envelope that delimits and unifies – a bark for their trunk, oxygen for their leaves, a living skin for their thoughts. That, if you don't mind, will be my last word. (APP 147)

This last word is a description of intelligent love, a capacity for mutual holding without illusion; but the penultimate word was still the desire to be not so much its subject as its object.

I want to return now to the fantasy of the common skin, which has hovered over the whole of this discussion. My view is, as adumbrated in chapter 3, that Anzieu has a number of different versions of the common skin, and each of these has a distinct relevance to the question of love. The first is the one on which Anzieu explicitly insists: "in mathematical terms, an interface, she on the one side, he [sic] on the other side of the same skin" (ACO 71) or "a more or less conscious fantasy of cutaneous fusion with the mother [...] in which the two bodies of the child and the mother have a common interface. The separation from the mother is conceived as the tearing of this common skin" (AMP 63). This fantasy includes – as we saw in the case of the "diameter" of the young couple – key elements of the future skin-ego, but it is only the preliminary stage towards a *moi-peau* since it is "linked to a symbiotic dependence between the two partners" (AEN 134), a kind of "adhesivity" (AEN 39) that can only lead to suffering.

However even this surface-to-surface version is less simple than it seems. The question of whether it is a 2D or 3D concept is raised by Albert Ciccone in the third part of *L'Épiderme nomade et la peau psychique* (AEN 134), and I should like to follow his lead in suggesting that we gain more by viewing the common skin as a relation within three dimensions – not in the manner of the Moebius ring but as something more like a Klein bottle, or a nucleus. This possibility is implied in the "homeostatic system" (AMP 81) which is both a fantasy and an actuality in good infant care: in a parallel with the "physical phenomenon of feed-back" (AMP 78, citing Brazelton) we may "consider the mother-nursling dyad as a single system formed out of interdependent elements communicating information between them, with the feed-back going in both directions, from mother to baby and baby to mother". As part of an "interactive spiral" (AEN 95) that will culminate in the skin-ego, this early version of active interchange, the 3D common skin, is very close to describing what goes on (fantasmatically in retrospection, really in prenatal life and in the life of the pregnant woman) in the womb. Here there literally is one skin in common: the mother's skin does double service, containing her inner body and that of the child, along with its skin to be sure but in the same way that her internal organs are surrounded by their own membranes.

I introduced this extension of the common skin in chapter 3, citing the passage in which there are "two beings in a single body [and Borges communicates the feeling of] one's own body and that other body from which one's own is not yet completely differentiated" (ACO 316). It is connected to the "Russian-dolls embeddedness [*emboîtement ... en poupées gigognes*]" (AEN 44) that we also find in Annie Anzieu's description of the psychoanalyst as intelligent mother. Or again: the womb is an "undifferentiated anatomo-psychic container [...], experienced as the bag that holds together the fragments of consciousness. The protection against excitation is constituted by the mother's body, particularly her belly. A field of sensitivity common to the foetus and the mother develops" (AEN 96). Later, even, the baby at the breast, held in its mother's arms, [re]experiences the sense of "its own skin contained by the skin of the mother" (AMP 219).

This uterine containment is totally reversible. In chapter 2 we saw "the group of children inside the womb of the mother, or that of the mother in the womb of the children" (AGI 108) and in chapter 1 Anzieu's determination to use psychoanalysis "to care for my mother in myself and other people" (APP 20). It is an essential preliminary to being able to interiorize the mother in a way that creates the skin-ego (AMP 121), the mind (AP 22, citing Winnicott) and the notion of space itself (AEN 49). In fact, psychic autonomy grows out of two inseparable "spheres" (AEN 93), that of primary narcissism and its other, "phantasized as the object containing all objects [...] the mother-space-everything", from which autonomy is a necessary escape.

This version of the common skin, combining the system of exchange and reversible containment of the sphere, is also cited as the idea of "reciprocal inclusion" or "mutual inclusion".[27] Like Anzieu, Sami-Ali sometimes describes this as a system of embedding like Russian dolls (SAE 43 and SAC 76) and it can be viewed as having only two dimensions (SAE 38) but it is essentially a projective effect, "a double process of sensorial and fantasmatic projection" (AMP 59) and thus, I would argue, a sphere rather than a surface-to-surface phenomenon.

With all this material from Anzieu, I'd like to return to the theories of love and the caress of our four philosophers and see how they marry up with his theory. To begin at the end: Irigaray's stress on maintaining the space of difference between lovers is close to Anzieu's "gap between abandonment to the other and abandonment of the other". She too acknowledges "the attraction of the hollow to the object" (AA 48) though she inverts its gender, as does Didier in his image of sex as a recollection of infant feeding: for her it is the male who demands to have his hollows filled up by the female. Most striking is the parallel between the skin-ego and Irigaray's "double envelope" which is necessary for the preservation of desire beyond the mucous. "Skin must protect the internal mucous world", she says. For Anzieu, skin and the mucous world are psychically and physically continuous.

As for Lévinas, there is little in his representation of the "blind appetite" of desire that finds echoes in Anzieu, other, perhaps, than in the latter's erotic fictions. In the teenage cousins exploring each other's surfaces, with the taboo attached only to the genitals and eyes, to prove both difference – "sex is not the essential difference between a man and a woman. This dissimilarity, which is also a complementarity, resides in the skin" (AEN 19) and exchangeability – "I saw myself [after the girl's death] double and entire, boy and girl, her and me" – gender is more central than anywhere else in his writings. What is more analogous is the use of the term "[umbilical] mystery" to describe the operation of responsibility for others, including friendship (ABe 248) or the "intelligent love" that protects and shields (APP 147).

Between Anzieu and Merleau-Ponty we find the reflexivity of the sense of touch feeding into a larger experience. "Tactile reflexivity", writes Anzieu, "provides the model for the later development of all the other sensorial reflexivities [and] based on this sensorial reflexivity, there develops the reflexivity of thought" (AEN 69). Add to this the potentiality of a world of "entwining" or "interlacing" and I suggest we find something like the "reciprocal inclusions" borrowed from Sami-Ali. Finally, there is a whole set of similarities in what I have taken as Merleau-Ponty's key representation of the caress. In "the body [with its whole expanse] tirelessly sculpting with its hands the strange statue which, in its turn, gives everything it receives" we surely find nothing so much as the creation of a psychic envelope of beauty. Then the embrace turns into a twin-like womb-like "bubble": "floating in Being together with another life, making itself the outside of its inside and the inside of its outside". What is this enclosure if not the second type of common skin?

But the most striking resemblances are with the theory of Sartre. Anzieu mentions Sartre rarely: the only extensive discussion is in his book on 1968 (ACI 78-82). Yet in the way they describe the love-relation, a startling number of parallels can be found. First of all, the creation of the other as flesh through the caress is, rather like Merleau-Ponty's aestheticizing version - and Sartre's term *faire naître* underlies it - a version of the creation of a psychic envelope which implicitly continues that of the other's infant environment. It is, like the common skin, a circuit - "The caress creates the other as flesh both for me and for themselves" (SEN 430) - and a mutual inclusion: a *"double reciprocal incarnation"* (431). And it is essentially a relation of passivity: here we recall Anzieu's stress on "being loved" (APP 147) rather than loving. For Sartre the "languor" of lovemaking, its reversal of the motive of grasping or taking hold - I will return to this in the next chapter - one's body "caressing itself against [that of the other] rather than caressing it", all these are the gestures of the wish to be loved: "it is not so much that we *take hold of* a part of the other's body but that we *bring* our own body up against the body of the other. Not so much pushing or touching, in an active sense, but *placing up against*".

The second passage I cited from Sartre goes further, again in a similar vein to Anzieu (but with anticipations also of our other three thinkers), for it describes love as a structuration of the world. The motif of passivity continues: each of the partners wishes to be the object for the other. In the same way, Anzieu writes of "the double belief that the partner is the object that counts above everything for me and that they also have the desire to be that primordial object for someone, who happens to be me" (ACD 254). This rather strange phrasing represents I think, the corollary of what I earlier called 'risk': the miracle, if it occurs, of two subjectivities coincidentally desiring one another. The object they each agree to be is "everything in the world" (SEN 408), and that means both the container of "all other *thises*" and the "limit-object of [the other's] transcendence [...] the circle of their freedom", with the other freely accepting this limitation as though it had always been so: "This shows us what the lover fundamentally wants from the beloved: he does not want to *act* upon the other's freedom but to exist *a priori* as the objective limit of that freedom".

I have likened this idyll to the common skin of mutual inclusion in which lovers, in their "bubble" (ACD 254), envelop each other. But actually it is something very much larger: not a membrane of enclosure (as, for example, in Lévinas's image) but one coterminous with the imaginable world. This fantasy is uterine, to be sure, a version of the "mother-space-everything" (AEN 93) and the combination of centre and circumference (AP 67); but it is also a projection. A projection must fall somewhere: this membrane is the screen or backcloth on which it falls. Sartre's term is *fond*, Anzieu speaks of a 'backcloth' or *toile de fond*. But again, let us remember, we are not talking about a two-dimensional backdrop but a membrane coextensive with the outer limit of the 'freedom' that is each individual's psychic world.

Remember the description of sexual pleasure as a repetition of infant feeding: this was cited to illustrate "the prototype of the relations between figure and ground" that found mental life (ACD 215). In *Le Moi-peau* we read of a story by Bioy Casares in which desire cinematically projects a lost world of love but the protagonist "can only die of it, for already his skin is beginning to fall away" (AMP 154); here, fatally, love and the real skin cannot coexist. Elsewhere too, the image of a backcloth is evoked. The skin-ego is the "backcloth" of the function of thought (AMP 175; APo 120); dream revives the envelope of inscription as its "screen, film [*pellicule*], backcloth" (AEN 103); but above all, this term refers to the sense of touch:

> In relation to all other sensorial registers, the tactile possesses a distinctive characteristic which not only makes it the origin of the psychic system but allows it to provide the latter permanently with something one could also call the mental background, the backcloth on which psychic contents are inscribed as figures, or perhaps the containing envelope which allows the psychic apparatus to be capable of having contents. (AMP 106-7)

The sense of touch is reversible, "bipolar" (AMP 107), creating internal and external perceptions – and it is the backcloth to everything else. So, when Sartre describes love thus: "I am the world. Instead of being a *this* standing out against a background of world, I am the object-background against which the world stands out" (409-10), he could be talking about the skin-ego as the site of consensuality – consensuality both in Anzieu's primary sense of a place where the senses coincide, but also in my sense, the sense of shared sensuality on a basis of love.

Chapter 9: Loss

To love, then, is to wish to be for the other the object that is the membrane of their world. In terms of the skin-ego, it is the wish to reproduce the other's original common skin, as a reciprocal inclusion, holding and being held.

I want to begin this chapter by looking at two films that, in very different ways, amplify this state of desire. They are as different, let us say, as Gautier's and Rilke's idea of the skin of the statue. The two films are Jane Campion's *The Piano* (1992) and Peter Weir's *The Truman* Show (1998). In the first, love creates a common skin through the deployment of that "expression" which is where Merleau-Ponty and Lévinas reached the end of their representation of the caress. In these theories (see chapter 8: 184 and 186) expression is impossible and paradoxical yet intrinsically tied to the caress. And this is what Anzieu said about the way the psychoanalyst may avoid the demand of touch in the analytic setting: "one can touch with the voice" (APP 75). He goes on:

> In everyday language we talk of 'getting in touch with someone' or 'having good contact' with them. This shows that the earliest version of contact was actually tactile, and then it was metaphorically transposed to the other sense organs and sensorial fields. What seems to me specifically analytical is to draw on these sense organs and sensorial fields in order to find verbal equivalents of those primal exchanges that introduce the nursling to the world of signifiers, but which have to be abandoned afterwards if the child is to develop.

By reversing this principle, *The Piano* shows us what happens when touch stands in for the voice. As in Merleau-Ponty, the caress in this film is an aestheticizing act, but one that changes the "strange statue" that Ada has been, along with her piano, into human flesh. This pygmalionesque process comes about, I'd like to suggest, not so much by her being caressed but by her own caressing.

At the beginning of *The Piano* we see fingers. A high-pitched woman's voice speaks lines that we sense are audible only to her and us. "The voice you hear is not my speaking voice but my mind's voice", she says, "I have not spoken since I was six years old". It is not clear that the hazy vertical stripes of reddish light are fingers until we see the reverse shot: a wedding-ringed hand with one dark eye looking out. Red tints are rare in the film, the most striking being the lacy cloth Baines uses to divide his hut, creating a feminized flavour in his male space. Aside from this, shades of sub-aqueous blue predominate, and flesh is very white. Preparing to leave Scotland, the protagonist tends to her sleeping daughter, from whom she cuts a roller-skate (echoed later in the shoe she looses off to save herself from drowning) and then turns to her piano:

> in the dim light she begins to play strongly. Her face strains, she is utterly involved, unaware of her own strange guttural sounds that form an eerie accompaniment to the music.
>
> An old maid in night-dress looks in. abruptly the woman stops playing. The emotion leaves her face, it whitens and seems solid like a wall. (Campion 1993: 10)

The guttural noises are omitted in the film, but the air of concentration and the "wall" are very clear. That wall is Ada's psychic skin. Annie Anzieu describes such silence in autistic mutes: "nothing emerges from [them] but nothing goes in either; their body is rather like an egg in motion. It seems as though they use this imaginary shell to protect an essential inner core" (APL 129). Ada is not autistic but she has made an autistic choice, "not a handicap but a strategy" (Campion in Pryor 25). All the characters think of her silence in the Victorian terms of an exorbitant 'will'. We are told when it began but not why or how (though see Pullinger's theory in Brace 11). Later we hear that she has played the piano since she was "five or six", but which came first and which may have caused or encouraged the other is never suggested.

If the voice is a normalizing emissary, Ada chooses two ways to replace its function. The first is her daughter (see Segal 1997), the second is the piano; both are also objects of intense intimacy and both are mediated by the hand. As we have already seen, when we watch Ada playing her piano we witness something that is inward-turned, involving all her senses together. Baines inherits this fascination with the piano as something that must be touched as well as heard; the blind piano-tuner also reads it with his fingers and his nose. It sends and refuses messages; it is awkward and beautiful. Thus it stands in for its owner's body and is an image of her relation to something of herself that does not form words. It is all the more essential as a means of communication between Ada and Baines because he cannot read so she cannot write.

Hands feature in most of the scenes of *The Piano*, not only when we hear the music. They flail and catch as Ada and Flora are uncomfortably transported onto the beach. They play, inseparably stroking and communicating, as the mother tells her daughter a story. They hold a portrait, a message, they need scrubbing, they dress and undress, they dictate rebuffs; they shadow-mime threat and carry out violence; they cover a mouth to indicate shock or frame a face viewing itself; they measure out a bargain or turn pages; a hand held in an audience may suggest marital intimacy (though Ada removes her ring to play the piano but not to make love); hands touch and caress.

What I am most interested in here is a particular use of the hand that occurs at a number of significant points in *The Piano*: the use of the back of the hand to caress a human or inanimate object. Ada's key objects are four: the piano, the child, the man and the sea. Before taking each of these in turn, I want to discuss the importance of the back of the hand. Rilke complained to Gide that

> the German language has a word for the back of the hand but no word for the palm.
>
> "At the most, we can say 'Handflächen': the flat [*plaine*] of the hand", he cried. "Imagine calling the inside of the hand a 'plain'! Whereas 'Handrücken' is used all the time. That's what they think about, the back of the hand, a surface with no interest, no personality, no sensuality, no softness, by contrast to the palm, which is warm, caressive, soft, and expresses all the mystery of the individual!" (Gide 1924: 61)

From the point of view of this film, he is entirely wrong. Eve Sedgwick recently wrote: "to touch is always already to reach out, to fondle, to heft, to tap or to enfold" (Sedgwick 2003: 14) but I disagree with her too. The touch of the back of a hand suggests the fingers in their least haptic, most patient mode; it is the obverse of grasping and very much what Sartre has in mind when he writes:

> my body becomes flesh in order to touch their body with its own passivity – in caressing itself against it rather than caressing it. This is why the gestures of lovemaking have a languor that one might almost call studied: it is not so much that we *take hold of* a part of the other's body but that we *bring* our own body up against the body of the other. Not so much pushing or touching, in an active sense, but *placing up against*. (SEN 431)

When Baines first asks Ada to come half-clothed to his bed, she lies face down, looking away, as though to protect herself against either his or her own desire. Later, as we shall see, Stewart feels more rather than less under threat when she makes him lie on his front. But between Ada and Baines her back has been the object of attention from the start; this has been his way to reach the woman making music who turns away towards her own playing, whose consensuality is self-directed – not, I think, because this is what refuses him nor because it exposes her "vulnerable spine" (Dyson 271) but because this inward-turning gesture is what is most fascinating about her. Thus he touches, kisses, gazes at the nape of her neck; but when he caresses it with the back of his hand, she scents danger and "changes the music to something brisk, almost comical" (Campion 1993: 60).

When, with the return of the piano, Ada loses this scenario, she discovers it is something she cannot do without. We see the changeover from her exclusive libidinous attachment to the instrument to a wish to be observed again by her lover when she sits down to play:

> She starts with wholehearted feeling, her eyes closed, but before long she is surprised by a moving reflection across the piano and she starts, glancing over her shoulder. She stops and begins again. But once more a reflex has her glance across her left shoulder and she pauses in her playing. Disquieted, she starts again and again she looks away. She stops, confused, unable to go on, unable to get up, one hand on the lid and one on the piano keys. (80)

At this point, then, she perceives that it is not enough to enjoy consensuality without the participation of another; she needs to be the object of a desiring gaze. This scene of playing is briefer in the film. We see Ada gaze without warmth at the piano while chewing a forkful of food; go over to it; place her napkin carelessly on it, and then pass the back of her hand across the keys, first one way and then the other. She starts to play, looks behind her, again runs the back of her fingers over the unsatisfying keys. This gesture has occurred once earlier on. With Baines kissing her shoulders and neck as she lies warily beside him, we watch her expression change from the backward-consciousness of him to a forward concern with the piano. He stops, she gets up, goes to the piano and runs the back of her hand along the keys – then he shuts the lid so that she has to pull her hand away, and with tears in her eyes, puts on her jacket and leaves.

In these scenes, the non-haptic, non-active relation of the fingers' back to the piano stands at the turning-point between two ways of living for Ada. The piano, like Baines, always faces her, offering itself front-on. By touching it through the back of her hand, she moves away from its prosthetic role in her address to herself and begins, uncertainly, to address herself to the other.

The second object that Ada touches with the back of her hand is her daughter. We see the girl take this initiative first, lying in bed laughing, caressing her mother's cheek with the back of her fingers while the latter signs to her the story of her relationship with Flora's father. Once, we see the back of Ada's hand on Flora: barricaded in Stewart's house, they are asleep together and the mother, apparently dreaming of her lover, passes the back of her hand across the child's night-dress and hair until this wakes her, shocked, and she moves on to her husband, in the scene we shall look at in a moment. The same placing, mother and daughter with hand-backs on each other's chest and neck, finds them asleep on the morning that Stewart decides to take down the planks. If it stands for anything, I suppose it is for innocence – which is the challenge we also take into the next example, Ada's caressing of her husband's body.

This is perhaps the most curious moment in *The Piano*, and one that Campion added as a supplement to the basic adultery tale. Her producer Jan Chapman comments: "I think this is the thing that makes the film modern actually, and not sentimental" (Campion 138) and Campion goes on:

> Ada actually uses her husband Stewart as a sexual object – this is the outrageous morality of the film – which seems very innocent but in fact has its power to be very surprising. I think many women have had the experience of feeling like a sexual object, and that's exactly what happens to Stewart. [...] to see a woman actually doing it, especially a Victorian woman, is somehow shocking – and to see a man so vulnerable. It becomes a relationship of power, the power of those that care and those that don't care. I'm very very interested in the brutal innocence of that. (138-39)

Sam Neill, who plays Stewart, several times the agent of a simpler brutal innocence, has this to add: "What happens to him, I think, is that this shell – a carapace that Victorian men could assume – is cracked and disintegrated by the power of his feelings for Ada, leaving him very exposed. I think of him as being a man who has lost all his skin" (147).

Yet when an almost sleepwalking Ada comes to his bed to caress him (having been, we surmise, shocked to find herself reaching for Flora while dreaming of Baines), it is his skin that she focuses on, caressing it carefully and seemingly tenderly, running the back of her hand across and around his face, turning it in his open palm. Bruzzi notes "the use of a luscious golden light and a fluid camera that (ironically in Stewart's case) intensifies the attraction of the male skin" (Bruzzi 261), and infers from it that Ada "has discovered a more abstract desire for closeness". Campion's stage-directions focus more on a strange equivalence of childlike qualities: Stewart "looks up into her face like a child after a bad dream, fearful and trusting" (Campion 89) but then "childlike she stops and kisses the soft skin of his belly. [...] Ada seems removed from Stewart as if she had a separate curiosity of her own" (89-90).

In the second scene, he is lying on his front. Ada has her eyes closed. She places first the front, then the back of her hand on his buttocks. When he leaps up, saying "I want to touch you. Why can't I touch you? Don't you like me?",[28] she just "looks back, moved by his helplessness, but distanced, as if it has nothing to do with her" (93). I will come back later in this chapter to the exposure and vulnerability of the eroticized skin that is not loved. What is interesting here is the dedication and glow of this desire redirected onto an unloved body and simply "curious" and detached. It is as much a surprise to see Stewart bathed in golden light as to see Baines draping a pink lace cloth in his hut, so dominant have the shades of blue been in this film.

It is with these shades of blue that my last example is concerned. The endgame has been played out, Stewart has 'heard' Ada's voice telling him to release her and he has 'returned' her to Baines. She is seated in the longboat that is carrying her, her lover, daughter, possessions and piano to Nelson. Suddenly she signs to the child that the piano must be thrown in the sea: "she doesn't want it; she says it's spoiled". At Ada's insistence Baines agrees. She strokes the water with the back of her hand – once, twice, as if preparing it or herself; then gazes at it. A moment later, she steps into the coiled rope and is pulled down with the piano.

Once again the director attributes her protagonist's action to "curiosity" (121), this time "fatal [...], odd and undisciplined". Moments later, of course, Ada chooses life and enters into the unexpected happy ending that many viewers cannot forgive. But my question is: why the back-of-the-hand caress at exactly this point? Is it because she is bidding a complicated and ambivalent farewell to the whole skin of that world in which she was consensual with herself, the "egg in motion" of her autistic enclosure, which is no longer in season now that missing parts can be connectives rather than only lacks?

A last few words on haptic touch. David Rodaway uses the term for the full range of the sense of touch (Rodaway 41-60) and Laura Marks defines it as "the combination of tactile, kinaesthetic and proprioceptive functions, the way we experience touch both on the surface of and inside our bodies" which she associates in video with "the caressing look" (Marks 2002: 2 and 8, borrowing from Deleuze and Guattari). But, leaning on its etymology in the Greek for 'fasten' or 'grasp' and its current usage in the science of cyber-touching (see Castañeda), I prefer to keep to a narrower definition that distinguishes it, as does Sartre, from the "studied languor" of the caress, which does not lay hold of anything. "I am always on the same side of my palm" writes Régine Detambel (123) and Rilke describes the palm as "warm, caressive, soft" and expressive (see also Josipovici 72); it is so when it is not grasping. As we have seen, the most searching expression of caressiveness in *The Piano* is enacted by the back of the hand.

There is a scene in *The Truman Show* that links it nicely to its writer-producer's earlier project *Gattaca*. It appears about halfway through the film, but in the self-referential 'teaser trailer' it is beamed out to audiences far and wide. Truman draws a space-helmet round his head on the surface of his bathroom mirror (actually a camera/screen) and hymns the planet of "Trumania of the Burbank Galaxy". The camera crew fear for a moment that he is looking straight out at them; but no, he is talking to himself. It is meant to indicate, of course, how childishly hopeful he remains, but it is also literally true. He and his world are indeed coextensive; yet one thing distinguishes them. "While the world he inhabits is in some respects counterfeit", says his creator Christof, "there's nothing fake about Truman himself".

In essence, the protagonist is the only true thing in the fallacious world that enwraps him. For this to be so, he must be the only one who is ignorant while everyone else alive knows what he does not, being either his minder or his audience. For him to inhabit a world geared exclusively to him, he must be the victim of a containing falsehood. Of course this turns out to be an inadequate skin-ego. But before that it is, in fantasy, the most complete skin-ego imaginable.

The American Dream inverted, this story is about forging through the membrane of the place in which you are tenderly loved. This love is not simple: we must agree that the cast who surround him – the production crew wearing T-shirts saying "Love him. Protect him", 'Meryl' shouting "how can anybody expect me to carry on under these conditions: it's not professional!", 'Marlon' jerking tears from Truman and the audience by repeating "the last thing I would ever do is lie to you" – are cynical; but the worldwide audience, the sort of people who "leave him on all night for comfort", they are the ones who sustain the membrane of love.

Truman's world has, thus, two membranes, just as Diana's had two or more circuits: the nearer one contains those within touching distance, who are or are not trustworthy, the further one the audience who imagine him – and whom, if he had knowledge, he would be able to imagine in return. But for him there is no circulation of this comforting air; his is a closed enclosure.

That is why Seahaven Island, unlike the sweet Pleasantville in which the fire service really do have nothing more to contend with than cats up trees, is a world girded with terror, a paranoid universe. All fictions are strictly paranoiac, founded on plot and subject to its laws; Truman's drama is that he learns he is living in such a theistic universe, bounded by the banal and avid imagination of Christof and the world audience. Their greed is his imprisonment. This is shown, half-comically, by the number of accidents and mishaps that intervene: a studio light crashes out of the sky at his feet, the rain falls only on him, a fire springs up in the middle of the road, cars or hospital trolleys crisscross to block his way, an emergency has occurred at the local nuclear plant, and all the posters at the travel bureau are meant to terrify.

Some of Truman's fears – dogs, the sea – have been carefully cultivated; he is, after all, the foetus that got out of the womb prematurely, the child that wanted to be an explorer. Others enclose him, not only when he moves towards the borders he must not cross, but also in the messages to us in the frequent use of point-of-view framing or oblique camera-angles. Still other hints, visible to him but comprehensible only to us, are present in the double-sunned sky or the sudden rise of day by brightness-and-shade crossing the townscape from above. In fact the presence of spherical objects is ubiquitous in this film: not only the five thousand cameras secreted around the set, most the shape of coloured dots, one spiralling into his office pencil-sharpener, but Truman's kitchen table and radio, desk tidies, hemispherical coffee-cups (which the old ladies have too, just as the man in the bath has a bowl full of round objects), the mini-roundabout he shoots round to frighten Meryl, the cherries on her uniform, the actors' loop, the capstan and other controls on the boat, even the door-knob saying 'Exit'. The gigantic studio, visible from space, is of course a hemisphere, as is Seahaven Island itself; and Christof, the "televisionary" responsible for the whole thing, is all circles from his bald patch to his beret and glasses. Different of course is Sylvia, who represents Truman as imprisoned behind bars: her clothing and décor emphasize the stripes that restrain rather than the globe that contains.

It is perhaps a little odd that the escape narrative is present from the very beginning of the film rather than following the establishment of a habit-world not yet subject to error and infiltration; perhaps the real audience needs leave to breathe from the very start, being able to believe in Truman by never seeing him content. We are, frustratingly or not, empowered to create a further membrane of love and knowledge into which we will guide the hero when he gets out from the intra-diegetic control of his fictional audience. When I referred, earlier, to the American Dream inverted, I meant that Truman begins where Vincent seems likely to end: a citizen of the personalized womb-world which is that hero's goal. The point he exemplifies is that you cannot have knowledge and desire at the same time and in the same place. What he wants, of course, is to be both known and loved by the same person and that is Sylvia.

Sylvia offers something different from the love of the audience because she has entered the first circle and known him both in the sense of having access to the truth and in the sense of recognizing him with her body. Their 'stolen' kiss seals a bargain which survives her disappearance in the form of the mythic Fiji he has to travel to. Where, actually, is she? We assume, somewhere in California, for she clatters down her stairs like Bonnie meeting Clyde when Truman has taken his step through the 'Exit' door. We do not see his happy-ever-after. The whole world cheers, though this means they have lost him, and turns to another channel; only the Omnicam corporation, his legal parent, will miss him. What this suggests is that there is a space between all the spherical membranes where Truman can be anonymous except to his beloved, a membrane made by love for its private purposes, where being the single subject of desire does not mean being the universal object of a plot.

Both *The Piano* and *The Truman Show* have happy endings dependent on an ideology of escape: Ada escapes from the shell of her skin, Truman from the shell of a controlled universe. Each is enabled to do this not only but also by the caressive touch and desiring gaze of another. Where they go from there is frozen in the films' close but it is suggested that they enter a sufficiently flexible bubble to survive successfully as couples. I want to pursue for a last moment the meaning of this touch-for-escape before turning to the time after such happiness is lost.

Robert Creeley's 1950s poem 'The Business' is, I believe, about taking a chance on love:[29]

> To be in love is like going out-
> side to see what kind of day
>
> it is. Do not
> mistake me. If you love
>
> her how prove she
> loves also, except that it
>
> occurs, a remote chance on
> which you stake
>
> yourself? But barter for
> the Indian was a means of sustenance.
>
> There are records.

Like Irigaray's concept of the caress as being the necessary step outside the "prison" (IPE 17), "house" (IPE 58) or "shell-shelter that protects you from an outside-you", this cautious poem proposes a balance of barter as the only basis for emotional sustenance. Truman and Ada take the same step. The point is that there are no guarantees. When you go outside to see what kind of day it is, it might well rain on you.

The rest of this chapter looks at loss, mourning and the skin of memory. The experience of loss is a temporal chain of formal signifiers, all spatial in relation to the subject's skin-ego: not who, what or why is the lost object, but *where* do we experience it in relation to our self? Classically, as I shall illustrate first, the lost object is thought of as introjected, incorporated or encrypted inside one's psychic body-space; but I shall also look for ways in which we might think about it as something we carry not within but beside or on top of us like an imaginary friend, a phantom limb or a second skin.

There are two kinds of lost love objects, the ones who disappear by death and the ones who disappear by betrayal. The mourning process for the two is similar in many ways, and it might well be argued that the latter is simply the former process with a preamble. Another way of comparing the two might be to observe that the ambivalence towards the lost object that follows betrayal is more overt than after a bereavement, just as, according to Freud, melancholia does unconsciously the work of ambivalence that mourning does consciously (see F-MM 266); so we might view the three processes as being on a continuum in which anger plays a diminishingly visible part. That is not to argue that any kind of mourning is less fraught or complex than any other; it is in the nature of loss to be deeply dramatic and complex and to haunt, as Proust shows, by little bites. The main point is that while both beloveds have left the orbit of the mourning subject, one still exists in the material elsewhere and can be imagined acting upon other desires; as we shall see later, this imagination affects the process of memory-formation and skin-formation. Here is an example of how one version of the lost object might, like a French train, be found hiding behind another.

In Victor Hugo's *Les Contemplations*, written mainly from exile in Guernsey, the collection of six 'books' is divided into two halves. Hugo commented: "Whoever only reads the first volume (*Autrefois* [*Long ago*]) will say 'It's all pink'. Whoever only reads the second volume (*Aujourd'hui* [*Nowadays*]) will say 'It's all black'" (Hugo xiii). We know that he exchanged real and pretended dates in order to put all his sad or lapidary poems into the second part coloured by the death of his elder daughter Léopoldine on her honeymoon in 1843. But a careful look reveals that the dividing line actually does not come at the point of her death; it comes at the point of her marriage. Two poems precede the famous line of dots representing the moment of her death. In other words, the loss that the father mourns is not so much a bereavement as a 'betrayal', her choice of another man to love in his place – and this may be why the most interesting of the poems carries a quite daring ambivalence.

In poem V: XXV, the lost object exists conserved and cannibalized in the poet's 'inside'. This is the place where poetry is made and, identified with the brain of Pluto, it makes the figure apostrophized in the first line, "O strophe du poète" [O stanza of the poet], an abducted Persephone transformed into language. Now it is the job of the poem to re-inter her so that she can never get back to the light except on this page. A verse is like a girl, abducted by force from a flowery meadow by "lui, le chercheur du gouffre obscur, le chasseur d'ombres" [him, searcher of the dark abyss, hunter of shadows] (Hugo 319) and held (like a Kleinian object surrounded by other objects):

> Prisonnière au plus noir de son âme profonde,
> Parmi les visions qui flottent comme l'onde,
> Sous son crâne à la fois céleste et souterrain,
> Assise, et t'accoudant sur un trône d'airain,
> Voyant dans ta mémoire, ainsi qu'une ombre vaine,
> Fuir l'éblouissement du jour et de la plaine,
> Par le maître gardée, et calme, et sans espoir,
> Tandis que, près de toi, les drames, groupe noir,
> Des sombres passions feuillettent le registre,
> Tu rêves dans sa nuit, Proserpine sinistre.
>   (Hugo 319-320)

> Captive in the blackest depths of his profound soul,
>   among visions drifting like waves
>   within his celestial and infernal skull,

> seated, leaning on a bronze throne,
> watching the bright light of day in the meadow fleeing
> in your memory like a vain shadow;
> guarded by the master, and calm, without hope,
> while close by you the black crowd of his dramas
> leaf through the register of dark passions,
> you dream in his night, sinister Proserpine.

The beloved preserved inside, this girl-verse, who will emerge again only as the Soylent Green of a published poem, has turned as sinister as her baleful abductor – why? because we do indeed digest our inner people with the sour juices of our psychic self. As Melanie Klein puts it in relation to the infant's introjection of the ambivalently loved mother: "the inside is felt to be a dangerous and poisonous place in which the loved object would perish" (Klein 1985a: 265). She is referring here to a psychic process involving two living people, but if we adapt it to the treatment of a dead beloved we can see how this entrapment by language is a double murder.

Something very similar happens in Romantic *récits*, in which the misloved woman – whose fault it obscurely is that the man could not love her properly – has to die so that the protagonist can turn narrator and not so much recover as re-inter her in text (see Segal 1988). A late avatar can be found in Gide: not only in his three best-known *récits L'Immoraliste* (1902), *La Porte étroite* (1909) and *La Symphonie pastorale* (1919), but also in his earliest and one of his last publications, both shaped by a direct autobiographical motive. The former is *Les Cahiers d'André Walter* [*The Notebooks of André Walter*] (1891), in which the eponymous writer is losing a race with his own protagonist as to who will go insane first. Full of biblical citation, guilt and undigested sexual fantasies, it was bizarrely written as a plea to Gide's cousin Madeleine to agree to marry him. The beloved in the fiction, familiarly called Emmanuèle, is persuaded by André's mother to marry another man and dies soon after. Losing her to marriage, he finds an interesting compensation, which concords at least somewhat with his fear of heterosexuality: "Bless you, dear mother, for above your deathbed our souls met" (Gide [1891, 1952] 1986: 87). Losing her to death, however, he triumphs: "She dies, *therefore* he possesses her... [...] Where do you exist now? only inside me: you live because I dream you, when I dream you and only then" (119 and 153).

The other text was written soon after Madeleine died in 1938 and published in 1951 after they were both dead (GJ3 1398). Its title, 'Et nunc manet in te', is taken from Virgil's account of Orpheus and Eurydice and means 'and now she remains [only] in you'. The elision of the subject of 'manet', the object in every sense of this homage, is apt enough, for there is an element of the same triumph in this text which, ostensibly intended to make reparation and revive the beloved, serves mainly to insist how much he truly loved her despite her dullness, anxiety and premature ageing. He certainly did grieve, but not in a straightforward way, as Maria Van Rysselberghe noted in April 1938:

> he has lost his counterweight, the fixed measure against which he tested his actions, his real tenderness, his greatest fidelity; in his inner dialogue, the other voice is silenced [...] I believe her memory will take on a firmer shape and, who knows, she may take up more room in his life than when she was alive. (CPD3 78)

The lost object conserved inside and reproduced in text may thus be sealed in another kind of tomb, the exposed tomb of literature where for a second time "she dies, *therefore* he possesses her". Why such punishment? Well for many reasons, but not least of these is the terrifying possibility that Eurydice no longer even knows who her mourner is. In Rilke's version, 'Orpheus. Eurydike. Hermes', she is sought in the labyrinth of "der Seelen wunderliches Bergwerk" [the soul's strange mine] but when she is found she is inaccessible in three ways: enclosed, as though pregnant with her own death:

> Sie war in sich, wie Eine hoher Hoffnung,
> und dachte nicht des Mannes, der voranging,
> und nicht des Weges, der ins Leben aufstieg.
> Sie war in sich. Und ihr Gestorbensein
> erfüllte sie wie Fülle.

> She was within herself, like one full with child,
> and thought nothing of the man walking ahead,
> nor of the path leading up to life.
> She was within herself. And her being-dead
> fulfilled her like abundance.

and at the same time "aufgelöst wie langes Haar" [dissolved like long hair] and rooted: "schon Wurzel" [already root]. This is exemplified when Orpheus turns round:

> Und als plötzlich jäh
> der Gott sie anhielt und mit Schmerz im Ausruf
> die Worte sprach: Er hat sich umgewendet -,
> begriff sie nichts und sagte leise: *Wer?*
>
> Fern aber, dunkel vor dem klaren Ausgang,
> stand irgend jemand, dessen Angesicht
> nicht zu erkennen war.
>
> And when suddenly, abruptly,
> the god stopped her and with painful voice
> said these words: "He has turned round" –
> she understood nothing and said softly *"Who?"*
>
> But far off, dark before the bright way-out,
> stood someone or other, whose countenance
> could not be recognized.

It is the nature of loss – whether to death or betrayal – that, reflexively, it destroys the loser as much as the one who has gone. Without the reciprocal gaze (the "counterweight") of the beloved, we may become simply "someone or other". Loss is weirdly contagious. As Anna Freud observes in a lighter vein (not talking of bereavement or betrayal but the child briefly mislaid in a department store): "It is interesting that children usually do not blame themselves for getting lost but instead blame the mother who lost them. An example of this was a little boy who, after being reunited with his mother, accused her tearfully, 'You losted me!' (not 'I lost you!')" (Anna Freud 101; see also AT 136).

The lost object conserved 'inside' is, as already noted, a staple of the psychoanalytic theory of mourning. I shall take a brief tour of the main theories, those of Freud, Klein and Abraham and Torok, before suggesting how these issues can be adapted to an idea of the object preserved 'outside'.

In 'Mourning and Melancholia' (1915, 1917), Freud insists that mourning is not pathological; though it involves a similar withdrawal from the world to melancholia, it is simply a slow and painful process of "reality-testing [... a] compromise by which the command of reality [acceptance that the beloved is gone] is carried out piecemeal" (F-MM 253; but also see Clewell 2004). Where mourning shades into melancholic pathology is where reality-testing is complicated by ambivalence. In spatial terms, the more conflictual the feelings for the lost person, the 'further down' - more embedded in the unconscious, less accessible to the preconscious and gradual dissipation - the process occurs: "by taking flight into the ego love escapes extinction" (267).

In the course of his discussion, Freud observes the shading of 'normal mourning' also into mania (see also RB 199-200), a point first brought to his attention by Karl Abraham, developed by Nicolas Abraham (no relation) and Maria Torok and, before them, Melanie Klein. Klein's essay on 'Mourning and its relation to manic-depressive states' (1940) argues that the process of getting through any kind of grief in later life repeats the one by which every child enters and lives through the depressive position in infancy. The infant has "incorporated his [sic] parents, feels them to be live people inside his body in the concrete way in which deep unconscious phantasies are experienced - they are, in his mind, 'internal' or 'inner' objects" (Klein 1985b: 345). Reality-testing at that stage - for after all the actual mother and father are probably present and alive - consists of a slightly risky checking in "the visible mother [for] continuous proofs of what the 'internal' mother is" like (346). These objects are coloured by extremes of energy or sorrow, goodness and badness. Triumph towards the internalized parents arises from the child's sense that it has reversed the power its parents have over it in the external world. In mourning, Klein argues, because "infantile death-wishes against parents, brothers and sisters are actually fulfilled whenever a loved person dies [...] early psychotic anxieties are reactivated" (354), so the bereaved person must work "to reinstate the lost love object in the ego" (353), both by reconstituting the good parental objects and by gradually finding a renewed "trust in external objects and values of various kinds" (355). This requires a renewed reality-testing, the re/acquisition of the ability to give vent to feelings, "projecting and ejecting" (359) the bad stuff. Eventually, as in Freud's scenario, "the love for the object wells up and the mourner feels more strongly that life *inside and outside* will go on after all and that the lost love object can be preserved within" (360, my italics).

The ambiguity of this last line is interesting. By "preserving the good object within", Klein sees the mourner as able to restore goodness outside as well – this refers, presumably, to other objects. She is only talking about bereavement, however; whereas the infant has a external mother to test against the internal one, the mourner is dealing with a good object that cannot change in 'reality' and thus has a certain stability, a stability that we might contrast with the fiancée done to death in *Les Cahiers d'André Walter* while her original was still very much around (and still refusing to marry him), as were the originals of the women in Constant's, Nerval's or Musset's equally murderous *récits*.

If internal objects take a lot of burying, this might help to explain the proliferation of good and bad terms for the process in the essays of Abraham and Torok's *L'Écorce et le noyau* [The Shell and the Kernel] ([1978] 1987). The familiar *introjection* (233, attributed to an article of 1909 by Sándor Ferenczi) comes to mean a healthy and productive process that may happen automatically and will in any case be aided by good psychoanalysis: "the process whereby the Unconscious is included in the Ego via objectal contacts. [...] the ultimate aim of introjection is not compensation but growth" (236). *Incorporation*, on the other hand, works by "taking possession of the object". The contrast is also one of time and realism: incorporation is a fantasy, "distinguished from introjection, a gradual process, by its instantaneous, magical character" (237). In sum, incorporation is a not part of effective mourning but a *"refusal of mourning"* (261), a refusal to acknowledge loss. There is a third version, "false incorporation" (264), to which I shall return presently; and a fourth, a "sort of anti-introjection [that] we have named *inclusion*" (297), wherein a memory is preserved by a process similar to the formation of a cocoon around a chrysalis, creating a "crypt" inside the ego which houses a family "phantom" (429): "the phantom is a formation of the unconscious which has the peculiarity of never having been conscious – for good reasons – and which results from being passed [...] from a parent's unconscious into a child's".

I am not disputing the centrality of the concept of inner reality in psychoanalysis, but rather (along with Anzieu) trying to examine the limits of insideness and explore a contrasting imaginative outside where things may be happening too. In this connection, then, I am going to retrace my steps backwards through these theorists to look for elements of the external function.

Abraham and Torok's theory is, of course, about shells as well as kernels. In fact in his 1968 essay 'L'écorce et le noyau', Abraham actually more often uses a term typical of Anzieu: 'envelope'. The envelope, identified by Freud with the ego, may be a protective ectoderm, but "it is marked by the thing it shelters, the thing that, concealed by it, is revealed in it" (205), the unconscious. The structure of Russian-dolls embedding continues throughout the system: "the Periphery itself includes a Kernel with its own Periphery which in turn includes a Kernel, and so on…" (217). At each level, the boundary faces two ways, and most significantly, memory is transmitted via "the messenger of the envelope" (218). Single in itself, like the mark on the mystic writing-pad, "the trace of an inscription can [...] lend itself to a double use: nucleic through the side turned towards the Unconscious and peripheric in its view towards the Conscious" – so memory traces "are situated on the nucleo-peripheric boundary" (219).

What follows is that the relation to objects situated outside is similar and symmetrical: "the originality of Freud [...] is that he rooted this intentional consciousness in a nucleo-peripheric system, thus giving it a fathomable depth, and the same is true of the Object, symmetrically the outside of the system" (220). The inner and the outer have a mirroring effect:

> It is through the symmetrical play of Object and Unconscious that we can recognize ourselves as the Object of the Object, with reciprocation of prerogatives: in other words, Consciousness is only possible through the Unconscious, whose image comes back to it via the Object. [...]
>
> Psychoanalytically, Consciousness is the organ of the Envelope, capable of objectifying the various modes of the nucleo-peripheric relation in the encounters of the Ego with external Objects. All psychoanalysis, clinical and theoretical, is based on this essential proposition.

There is, thus, in Abraham and Torok's writing, no object within that is not part of a continuous system reaching to the outside. We might expect less fluidity in Melanie Klein's theory but there is a moment of interesting indeterminacy in her essay on mourning. In relation to a dreamer she is citing as an example, she notes: "I have often found that processes which the patient unconsciously feels are going on inside him are represented as something happening on top of or closely round him" (Klein 1985b: 365). But two pages later this is swept away: "An attack on the outside of the body often stands for one which is felt to happen internally. I have already pointed out that something represented as being on top of or tightly round the body often has the deeper meaning of being inside" (367). Together with the ambiguity of 'feeling' in both these statements, the phrase "deeper meaning" seems to preempt what it needs to prove.

In Freud's essay the outside is present particularly in the basic contrast he draws between melancholia or mania, which are pathological versions of 'normal' mourning, and the latter which is the ability to submit a loss to the test of external reality. In reference to normal mourning, Freud says little about introjection. Even a phrase he uses about melancholia, in fact, suggests something more external than internal: "The complex of melancholia behaves like an open wound, drawing to itself cathectic energies" (F-MM 262); an open wound can only be on the surface of a body, facing outwards. It is strange, then, that Abraham and Torok, borrowing this "recurrent image" (272), somewhat perversely turn it inside out: "this is the wound that the melancholic tries to disguise, to surround with a wall, to encrypt".

The 'way out' is thus available between the lines in all three of these theories, and I should like to follow it by looking at some images of this kind of 'external' lost object. First: is the lost object an 'imaginary friend'?

Recall the comment of Adam Mars-Jones on the recently-dead Diana: "If celebrities are, as one theory has it, the equivalent in adults' lives of children's imaginary friends, then how does an imaginary friend go about making real ones?". This is Truman Burbank's problem; but the people who think they love him do not struggle with the same enigma. They can leave the TV on all night to hear him sleep, and if they feel sentimental once he has left the screen they can watch repeats. Judith Williamson was highlighting the fallacy in this kind of memorialization when she wrote: "how much easier it is to pour out all that intensity to someone we didn't know, *who's not there* and, in a sense, never was".

All this implies that grief should properly be measured against the 'reality' of the loss; if the latter does not count, it is merely "virtual grief" (Levy 33), however palpable the tears or feelings. But isn't all grief virtual, in a way? Aren't we, in every case, dealing with the ghost of someone whose presence has now been drastically problematized – all the more in betrayal, but in bereavement too – by our ambivalence and their inability or refusal to provide any reality-testing of a positive kind? All we test when, as Freud rather cheerfully notes, we gradually accept loss, is the absence of the beloved's existence in real space around us. We can, if we wish, use this imaginary friend fantasy in a positive way: friendly ghosts, risen messiahs or guardian angels. Here is one, from Eliot's *The Waste Land*:

> Who is the third who walks always beside you?
> When I count, there are only you and I together
> But when I look ahead up the white road
> There is always another one walking beside you

> Gliding wrapt in a brown mantle, hooded
> I do not know whether a man or a woman
> – But who is that on the other side of you?

Another leans down thoughtfully in Wim Wenders' *Der Himmel über Berlin* [*Wings of Desire*, 1987]. Angels are the internal externalized, as are other creatures of faith – but they are not what Klein called "on top of or closely round" you. At best they might seem to "walk beside" you; more often, they lean from afar or hover, like the Flaubertian God; the nearest they might get to you is behind you, where you cannot see them, like a psychoanalyst. Rilke's angels are entirely preoccupied with themselves and traditional angels are by definition the messengers of the divine. Flora carries a blighted message still dressed in her party wings. As a fantasy, an angel might aid and abet you, but it does not emanate from you as your imaginary friend does.

> Binker – what I call him – is a secret of my own,
> And Binker is the reason why I never feel alone.
> Playing in the nursery, sitting on the stair,
> Whatever I am busy at, Binker will be there.
> (A. A. Milne, *Now we are six*, 1927)

An imaginary friend is the interlocutor in a dialogue you reliably write. Less haunting than the replacement child's lost other, they eat up your mental scraps – or allow you, like Binker or the example cited below, to get double helpings. They have a partial relation to desire, perhaps because they too easily fulfil it, and the desires they answer are childish ones, like the wishes fulfilled by children's dreams. In Edith Olivier's *The Love Child* (also 1927) a bereaved single woman recalls an imaginary friend whom she was made to give up at the age of fourteen: "as the old memory came back, it seemed to Agatha that in losing Clarissa, she had not only lost a real playmate, but she had also lost the only being who had ever awoken her own personality, and made it responsive" (Olivier 16). Clarissa appears, is passed off as an adopted daughter and stays for several years. She vanishes, though, when adolescence brings sexuality, both because she desires and because she is heterosexually desired: she is no longer the emanation only of the protagonist but has called up an alternative identity, in a young man's fantasy.

Abraham and Torok's "false incorporation" leads to another example of the use of the imaginary friend by someone suffering bereavement:

> a man seated alone at a restaurant table who ordered two different meals at the same time; he ate both of them, on his own, as though he were sitting with another person. This man was visibly hallucinating the presence of a dead loved one but clearly he had not had to resort to incorporation. Quite the contrary, we can surmise: by means of this 'shared' meal he could keep the other outside the limits of his body, while filling the emptiness of his mouth, without having to 'absorb' the lost person. He seemed to be telling himself: "no, my loved one is not dead, s/he is still here, like before, with the same tastes, choosing the same favourite foods". (AT 264-65)

In both these cases, the imaginary friend serves a prosthetic function. Hallucinated or not – and in adults we tend to insist that if it is not fiction it must be madness – the creature is both of the self and a whole being separate from it. Agatha recovers through Clarissa the ability to play "with all the zest and spirit of her own childhood, and yet there was something added" (Olivier 25). Once bodily manifest, the child has a distinct and resistant personality and quickly becomes visible to other people; the invisible loved one of the man in the restaurant likes different dishes from him; Binker and the angels serve their real companions by supporting their wishes as a prosthesis serves the person who uses it (see Smith and Morra), but they are projected outwards as fully realized bodies.

Or is the lost object something more like a phantom limb? "Most people know what is meant", writes the neurologist Vilayanur Ramachandran, "by a phantom limb. A patient has an arm amputated because it has a malignant tumour or has been irreparably damaged in an accident but continues to feel the presence of the amputated arm" (Ramachandran 10). First named by Silas Weir Mitchell in 1872, it had already been recognized by Ambroise

Paré in 1649; a famous example is Lord Nelson who identified the twinges he felt in his lost arm as "direct evidence for he existence of the soul" (RB 22). The feeling might be tingling, clenching or twisting of fingers or toes – but the most common type of feeling in the phantom limb is pain.

Unlike an imaginary friend, the phantom limb is a externally projected part-object and, as with the lost love object, it is hard to know what exactly is meant by saying 'it hurts'. The impersonal 'it' in 'it's raining or 'il pleut' is said to refer to God or Jupiter. But the pain I might feel in a phantom limb is more like the sense of combined grief and accusation that Anna Freud's child patient felt on behalf of the mother who let him get lost. We always have this mixture of ownership and repudiation in relation to body parts that plague us, but in specific relation to the phantom limb this is a bereaved relation. As Elizabeth Grosz puts it: "The phantom is an expression of nostalgia for the unity and wholeness of the body, its completion. It is a memorial to the missing limb, a psychical delegate that stands in its place" (Grosz 73).

Phantom limbs may not only be experienced negatively. Like imaginary friends, they have their uses. Oliver Sacks points out how they can be essential to recovery:

> All amputees, and all who work with them, know that a phantom limb is essential if an artificial limb is to be used. Dr Michael Kremer writes: "Its value to the amputee is enormous. I am quite certain that no amputee with an artificial lower limb can walk on it satisfactorily until the body-image, in other words the phantom, is incorporated into it."
>
> Thus the disappearance of a phantom may be disastrous, and its recovery, its re-animation, a matter of urgency. (Sacks 1986: 64)

Abraham and Torok (before they adopted it as the word for the ghost-like effect of the encrypted secret) use the term 'phantom' to mean a phantom limb. Here they describe a use made of the lost other as prosthetic phantom:

> When I am melancholic I create the show of excessive mourning by staging the pain of the object that has lost me.
>
> Melancholics seem to torment their own flesh by lending their suffering to the body of their phantom; this has been seen [by Freud] as aggression turned back on the self. We do not know if they love their phantom but it is clear that the latter is 'crazy' about them, would do anything for them. The melancholic embodies the phantom in all the things that the desperate phantom would do 'for them'. (AT 274)

In this way the phantom carries the pain of the sufferer. It also "haunts the counter-transference", for "in order to objectalize the aggression the analyst will often target the phantom object" rather than the patient. So here, again, reciprocation, like a blood flow into and out of a real limb, binds the non-existent thing to me.

Pain in the phantom limb is, of course, most commonly thought of as problematic: in psychological terms, as frustrating the adaptive work of mourning rather than facilitating it. Ramachandran has brought clinical relief by the use of the "virtual reality box" (RB 46). This is a device that uses a mirror to delude the brain into thinking the lost hand is really there: the person wriggles right-hand fingers and the absent, painfully cramped left-hand fingers appear to wriggle back, which miraculously releases the tension in the phantom. In this way, as he puts it, he achieved "probably the first example in medical history of the successful 'amputation' of a phantom limb!" (49). A fictional realization of the imagined other thus, as in the case of Agatha's Clarissa, produces an effect of false – but effective – reassurance.

The possibilities of the phantom metaphor go further than one might think. You do not have to have had an amputation in order to have a phantom limb. Ramachandran cites the case of a woman born without arms who describes vivid sensations, including the fact that her phantom arms are "'about six to eight inches too short'" (RB 41). Conversely there is the case of 'supernumerary' or 'anarchic limb'. At a conference on *Phantom limb phenomena: A neurobiological diagnosis with aesthetic, cultural and philosophic implications*, run by artist Warren Neidich at Goldsmiths College

London in January 2005, Dave McGonigle cited the case of a woman with two arms and two legs who was convinced she had a third one of each at the left side of her body. She had to give up her job in a fish-sorting factory because the 'extra arm' kept getting in the way; it also made her avoid supermarkets because she felt it was liable to shoplift. Both these imagined additions to the body make the person actively more than they otherwise were. Even more thought-provoking, at the same conference, Chris Frith said: "We all have phantom limbs except that most of the time our real limbs coincide with them".

Is our whole body a phantom limb? Are we all our own Doppelgänger? If we take Frith's remark and apply it to the lost object, what it suggests is that our 'real' objects are as lost as the lost ones, and that not only is all grief 'virtual' but all other emotions are too. This leads to another kind of failed recognition. In a case viewed twice by Oliver Sacks (Sacks 1984 and 1985; see also Grosz 89), a person believes his leg to be a dead limb that pranksters have put in his bed; when he throws it out in horror he finds himself on the floor. Ramachandran cites the similar case of a woman who points to her own paralysed arm and says it is her brother's (RB 131). What is it that produces such alienation from something that has hitherto been uncomplicatedly part of oneself? Two further aberrant conditions may shed some light on this.

A person with Capgras' syndrome is able to recognize their mother or husband as looking exactly like that relative but they fail to feel the normal response to them and thus take them for an impostor. Ramachandran discusses the case of a young man called Arthur who, after a car accident, could no longer relate emotionally to his parents. He had no problems with them on the phone, and briefly he was convinced by being told the false father had gone to China, but there was no return of the global ability to love them on sight. Ramachandran makes two connections that are of especial interest here. The first is in relation to "the mechanics of how we form memories, in particular our ability to create enduring representations of faces" (169). Arthur is unable to see that three photos of the same person looking straight-on, slightly to one side and a little further averted are the same face – "only when the model's eyes were looking way off to one side was he able to discern correctly that she wasn't looking at him" (168) – for, somewhat like Leonard in *Memento* or patient H. M who has had both hippocampi removed (RB 15 and 149), he fails to form new memories of people. The second is that Capgras' syndrome can be connected to another syndrome called Cotard's, in which "a patient will assert that he is dead" (167). Arthur says to his mother one day: "'if the real Arthur ever returns, do you promise that you will still treat me as a friend and love me?'" (172).

If, as Ramachandran suggests, "Cotard's is simply an exaggerated form of Capgras' syndrome" (167), then we see again how contagious loss is. We have found this in a number of contexts: in Eurydice whose unrecognition turns Orpheus into "someone or other", in the lost child who accuses his mother, in the melancholic's projection of aggression onto the phantom other. This is, I suggest, the basic mechanism of the externalization of loss; it is actually an imagined encounter in which both parties are metamorphosed. After a betrayal we know the person who once loved us is still alive; we might even see them from time to time. But when, like Arthur or Eurydice, they fail to recognize us with love – when, that is, they manifest Capgras's syndrome towards us – then we become *their* phantom limb.

And what of the second skin? Missing Albertine – first by betrayal, then, without seeing her again, by bereavement – Proust's Marcel recalls the skin-like effects of both love and loss: "in despair I thought of that integument of caresses, kisses, companionable nights, which soon I would have to let myself be stripped of for ever [...] Indeed the memory of my desires was just as saturated with her, and with suffering, as the memory of my pleasures" (Proust [1925] 1954b, 482-83). By "desires" here he means the things he wanted but she did not fulfil when she was there. It is because she was the skin both of his desires and pleasures and also of the feeling of frustration at being prevented by monogamy from seeking other pleasures that, paradoxically or not, she is now an obstacle to finding those other things: "Albertine had seemed an obstacle between me and all things, because to me she was their container and I received them from her as from the vessel that held them. Now that vessel was broken, I no longer had the courage to seize them" (438). This exemplifies how the love-object is the skin of our experience of the world: as well as holding or containing us, that skin also shapes our ability to feel. Once it is lost, our hand cannot touch or reach.

If love enwraps, loss flays. Another version of the containing effect of love is the skin that is anointed by what I called earlier the desiring gaze. Different but inseparable from the touch, we have seen how it alters the body schema of Ada as she plays, extending it outward as if by a second skin she carried on her back. Consensuality moves from autistic to caressive mode. She transfers this *faute de mieux* to the skin of Stewart which she treats both "curiously" and "as if it has nothing to do with her". The effect on him of her refusal of feeling is to make him "a man who has lost all his skin".

Thus the loss of love strips away the top layer of skin, the one that glows. In chapter 5 I contrasted the "ideal circuit of intimate attention" with the "much larger circuit [that Diana had] with the anonymous multitudes". The latter is one expedient; another is the melancholic, autistic response of Ada (presumably) after the loss of her child's father; both are new skins that are 'grown' with time to scar over the lost surface. But I am interested here in something that never quite grafts on, something more like a garment, not overlaid but worn, and which preserves what has been lost.

When his beloved cousin dies, the protagonist of Anzieu's story 'L'épiderme nomade' carries out

> the work of mourning. I reverted to the fantasies and rituals of my childhood. I could not go to sleep without imagining my beloved dying in our embraces, and me laying down her body and stripping it of its skin, which I processed with the assistance of an embalmer. As long as this skin was available to my thought, my lost love was alive. In the morning the only way I could make myself get out of bed was by imagining putting on that skin. I looked at myself in the long mirror in the hall and saw myself double and complete: boy and girl, her and me. That is how I was able to go on living. (ACon 225; AEN 19).

After a lifetime of trading in cosmetic grafts, "cultivating the skin as a work of art as well as a means of primal communication between humans" (ACon

227; AEN 21), he prepares to repeat that eroticized consolatory gesture by creating his own shroud: "this second, incorruptible skin drawn from the multitude of people I have known, who will always go with me. This cloak of suppleness, beauty and warmth will wrap me in its illusion for the long passage of eternity" (ACon 233; AEN 26).

We will see another patchwork cloak made of skin presently. For the moment, let us remember how common it is for a bereaved person literally to keep and wear things that belonged to the lost one: two dressing-gowns remaining side by side on the door, Kathryn Harrison's inmixing of her mother's clothes and make-up with her own (Harrison 2005: 53) or Anna Freud wrapping herself in her father's old Loden coat during her own last days (Young-Bruehl 453). As in my other examples, this both covers and conserves the skin of loss.

In *Le Corps de l'œuvre*, Anzieu links together three types of psychic work:

> Dream, mourning, creation: what they have in common is that they are all phases of crisis for the psychic apparatus [...] Like the work of mourning, [creation] struggles with lack, loss, exile, pain; it incarnates an identification with the loved object that is gone, for instance, in the shape of the characters of the novel; it awakens both dormant areas of the libido and the self-destructive drives. (19-20)

Creation is, of course, a traditional skin for loss, allowing the mourner to go beyond it; it is the 'elegiac' impulse critiqued by Tammy Clewell, who argues, on the contrary, for "anticonsolatory and anti-idealist mourning practices that have gained widespread currency in the post-World War II era" (Clewell 2005: 57). In one sense, all texts are palimpsestic parchments. But I want to argue that Anzieu's theory is, much more interestingly, a meditation on the way the skin keeps alive and safe the memory of what is lost. His Marsyas, like that of Valverde, carries his flayed skin with a certain grace. In earlier chapters I have shown how the inside-outside relation in Anzieu's theory adapts itself to what Abraham and Torok would call the 'nucleo-peripheric' structure, exemplified in the common skin with the mother and its later avatar, the mutual containment of lovers. I want to bring these together now by turning now again to the replacement child.

I have referred earlier in this book to a number of children who, in different ways, replaced older siblings of the same sex or the other, whose death either preceded or followed their birth. Here is one I have not mentioned before:

> I lived through my death before living my life. My brother had died of meningitis at the age of seven, three years before I was born. This experience shook my mother to her very depths. Everything about my brother – his brilliance, his early genius, grace and beauty – had been a thrill to her. His death was a terrible shock. She never got over it. Not until my birth was my parents' despair somewhat lightened, but even then their misfortune permeated every cell of their bodies. I could already feel their anguish in my mother's womb. My foetus bathed in an infernal placenta [sic]. This anguish has never left me. How often have I recalled the life and death of this older brother, whose souvenirs were everywhere from the moment I was aware of anything – clothes, pictures, toys – and who had left indelible emotional traces in my parents' memories. I felt his persisting presence very deeply: both as a trauma – robbing me of affection – and as a passionate wish to outdo him. From then on all my efforts were dedicated to winning back my right to life, first and foremost by provoking the perpetual attention and interest of my family by a sort of constant aggressiveness. [...]
> My brother had lived for seven years. I think of him as a trial-run for myself, a sort of extreme genius. His brain had blown like an electric circuit overheated by incredible precocity. It was no accident that he was named Salvador, like my father, Salvador Dali i Cusi, and like me. He was the beloved one; I was over-loved. When I was born, I placed my feet in the

steps of the adored dead one, whom they went on loving through me, perhaps even more than before. (Dalí/Parinaud 12-13)

Dalí replaced a dead child of the same sex and after whom he was named, as did Marguerite Anzieu. J. M. Barrie, whose situation was otherwise rather similar, lost a brother during his lifetime. He was born in 1860 in the village of Kirriemuir in Scotland; his father was a self-employed weaver, and both parents were ambitious for the education of their sons. James was the third son and ninth birth of ten, following the deaths of two sisters. His mother, Margaret Ogilvy, whose biography he wrote in 1896, had herself lost her mother at the age of eight and thereafter, like Wendy, played mother to her younger brother and father. Barrie wrote of her:

> The reason why my books deal with the past instead of with the life I myself have known is simply this, I soon grow tired of writing tales unless I can see a little girl, of whom my mother has told me, wandering confidently through the pages. Such a grip has her memory of her girlhood had upon me since the age of six. (cited Dunbar 24; see also Bowlby 1998: 164-65)

The age of six is not insignificant. This was his age when his older brother, his mother's favourite and named David after both her brother and husband, was killed in a skating accident just before his fourteenth birthday. In her remaining twenty-nine years she never got over the loss and James's childhood seems to have been shaped by the attempt to make it up to her, trying to imitate his older brother's swagger, stance and whistle. In this he was both, obviously, trying to be loved for himself and accepting that he could only be the failed substitute for another. He tells how, a few days after the death, his older sister Jane Ann told him to go to his mother and "say to her that she still had another boy" (cited Birkin 4). After a silence,

> "I heard a listless voice that had never been listless before say. 'Is that you? I think the tone hurt me, for I made no answer, and then the voice said more anxiously 'Is that you?

again. I thought it was the dead boy she was speaking to, and I said in a little lonely voice, 'No, it's no' him, it's just me.' Then I heard a cry, and my mother turned in bed, and though it was dark I knew that she was holding out her arms.

After that I sat a great deal in her bed trying to make her forget him. [...] At first, they say, I was often jealous, stopping her fond memories with the cry, 'Do you mind nothing about me?' but that did not last; its place was taken by an intense desire [...] to become so like him that even my mother should not see the difference, and many and artful were the questions I put to that end. Then I practised in secret, but after a whole week had passed I was still rather like myself". (cited Birkin 4-5)

He never grew taller than a bit over 5': this too contributes to the almost seamless image of the 'boy who never grew up'. But again, the issue is rather more complex. Barrie himself did, if not grow up, at least grow old: "when I became a man", he wrote aged thirty-six and referring of course to his brother "he was still a boy of thirteen" (cited Birkin 5). His writing is full of lost boys, especially *Peter Pan and Wendy* (1911) and its fascinating forerunner, *The Little White Bird* (1902). "'Long ago'", Peter tells Wendy, "'I thought like you that my mother would always keep the window open for me; so I stayed away for moons and moons and moons, and then flew back; but the window was barred, for mother had forgotten all about me, and there was another little boy sleeping in my bed'" (Barrie [1911] 1988: 102). The most interesting thing about this fantasy is its doubled revenge: he is telling the story (for her sake?) from the viewpoint of the lost elder brother trying to return, while his younger self is the substitute in the mother's arms for whom window-bars are put on to exclude any rival old or new.

Rilke was also a replacement child. He was born prematurely in 1875 to parents who had lost a daughter soon after birth the year before. The name Rainer was given to him only years later by Lou Andreas-Salomé; it is a reasonable guess that his baptismal name René is at least partly in recognition of the child he replaced. As I pointed out earlier, his hatred of dolls seems to be rooted in the seductive-disturbing experience of being dressed up in girls' clothes, petted and passed around, which he describes in later letters (Freedman 10, Leppmann 13, Prater 5). The myth of his mother Phia is of a socially frustrated and ambitious middle-class woman who had been brought up in opulence and who cultivated the fantasies of aristocratic provenance that remained with her son, helping to keep him well supplied with wealthy patrons. We cannot know, as with Marguerite Anzieu, how far her mystical or creative fantasies betoken real gifts (see Leppmann 7-8), but she remains as fascinating as resented in Rilke's writing. In the third Elegy, he describes a mother's protection against stimuli:

> Mutter, *du* machtest ihn klein, du warsts, die ihn anfing;
> dir war er neu, du beugtest über die neuen
> Augen die freundliche Welt und wehrtest der fremden.

> Mother, *you* made him small, it was you who began him;
> he was new for you, who brought down to his new
> eyes the friendly world and kept away dangers.

But in the fifth, written nineteen years later, the boy *saltimbanque* desperately darts a "sweet countenance across to" his "rarely tender mother". Similarly, in *Die Aufzeichnungen des Malte Laurids Brigge* (1910), the mother is more absent than present – "Maman never came in the night; or rather yes, once she did" (RMLB 92) – yet the world of her family, full of ghosts, corridors, lost loves, houses half-burnt and intimations of madness, is powerfully evoked.

The adult Malte recalls shared times with his mother, gazing together at hanks of creamy lace, talk of the death of his sister who expired with childish resignation saying: "'I don't want to any more'" (RMLB 82) and her reappearance as a ghost only the dog could see. Immediately after the narrative of the pretend Sophie, he tells of a day when he wandered into the attic and found a row of wardrobes full of old garments. What most enchants him are "the capacious mantles, the cloths, the shawls, the veils, all those yielding, large, unused fabrics, that were so soft and caressive [*weich und schmeichelnd*]" (99) and the masks with bizarre expressions and real beards and eyebrows. But seduction turns to fear when, looking at himself dressed up, in a mirror made of green glass fragments, he knocks over a little table of ornaments, breaking a small bottle of scent, and suddenly he no longer knows who he is:

> That was just what the mirror was waiting for. Its moment of revenge had come. As I struggled, in a state of boundlessly increasing anxiety, to squeeze myself somehow out of my disguise, it forced me, I have no idea how, to look up and imposed on me an image – no, a reality, an alien, unbelievable, monstrous reality, which permeated me against my will. Now it was the stronger one and I was the mirror. I stared at this great, terrifying stranger in front of me and it seemed appalling to be alone with him. But at the very moment I thought that, something even worse happened: I lost all sense, I simply ceased to exist. For one second I had an indescribable, sad, and futile longing for myself, then there was only him – there was nothing but him. (101-02).

He runs away "but now it was [the other] that was running" (102), and when he finds his way back to familiar people, they think it is just part of the performance and laugh as he begs to be released. Finally they realize he has fainted and is lying there in all the wrappings "just like a log" [*rein wie ein Stück*] (103). Here (as elsewhere in this book), the masks and skins of the indeterminate other penetrate, invade and stick to the unanchored self.

In these examples, however different the circumstances, two elements predominate: the sense of a bitter obligation to the bereaved mother and the stress on gender uncertainty. Rilke identifies with an excessive and sartorial femininity which later he repudiates in the shape of the despised dolls; Marguerite Anzieu is attracted to "the male soul; Dalí affects a "constant aggressiveness"; Barrie swaggers, whistles and later creates the "cocky" (Barrie [1911] 1988: 13), "lovely" (16), "elegant" (37), "careless" (44) Peter Pan. A curious case featured in the tabloid magazine *Now* in 2006 is that of 'Paula' Rowe, who says:

> My two-year-old sister died before I was born. Mum had another four boys, but always wanted another girl. Part of me wonders if changing gender [sic] was my way of trying to give her the second daughter she longed for. For years I occasionally dressed as a woman to relieve stress. My first wife accepted it and would do my hair in pigtails, but my second wife never knew.
>
> In 1986 I suffered a series of crises. Mum died, my second marriage broke up and I was made redundant. Becoming Paula was an alternative to suicide. Losing my job felt like losing my masculinity. I'd have done anything to my body in order to change my life.
>
> [...] Having become Paula I felt like an actor in a play that never ends. In 1990 I met a man called Peter who was lonely. He fell in love with my personality and accepted me for who I was. [Since Peter's death in 1992,] today I'm not really a woman, but I'm not a man either. I feel guilty whichever public toilet I use. When I first transitioned, my friends called me Dolly Parton because my breasts are size 44D. They get in the way, but I've decided not to have reduction surgery. I don't want to cut myself any more. There's been enough pain in my life. (Rowe 33)

Overdetermined motives here familiarly include the needy mother, the excessive femininity, the insecure masculinity; they also lead to an indeterminate state we could call 'haunted'. A caption reads: "Today Paul covers his breasts with baggy clothes and no longer wears make-up": thus here too a garment of the lost self covers a post-altered body (for a full discussion of transsexuality and the skin, see Prosser).

In a dream, Anzieu's patient 'Palatine' is visited by the daughter of a well-known musician who has recently died. The girl shows her an object she is keeping in memory of him: "it is a text in the form of a skin, or rather several parchment skins attached together by rough oversewn stitches and covered in a very legible ink" (AGI 241). She is not allowed to touch it without gloves. Anzieu connects this to his patient's work as the editor of a selection of twentieth-century poetry. It becomes clear

> that the patchwork of skins also represented the image of Palatine's family body. An only child born after a dead brother, she was a piece of replacement in the family skin. She had been brought up at home by three women: her mother, the latter's sister and the mother of both (Palatine's grandmother). [...] She adored her father, the only man of the family, but the three women vigilantly kept them apart, forming a screen between him and her. The family skin constituted by these three women and inherited from the previous generation held the father at bay – he was after all from outside – and became the family skin for the new family. After we had interpreted her dream and she had absorbed its discoveries, Palatine no longer experienced herself as a 'selection piece' chosen [*morceau choisi*] by others, but as a whole that she chose to be. (AGI 242)

Once she is able to recognize this, she is able not only to experience her family context globally but to separate from it without terror. This case history brings together the fate of the replacement child, with its gender imbalances (the girl is shown as trying to regain contact with male figures held at bay), with the garment worn in memory of someone lost. Palatine has been "a piece of replacement in the family skin"; now, by bringing this dream to this psychoanalyst, she acquires a "family skin" in whose illusion she can comfortably "wrap herself" for the rest of her life. The opposite of the crypts of Abraham and Torok, this consciously externalized skin of memory can serve its possessor.

The choice of having a replacement child is, writes Andrea Sabbadini, sometimes "a pathological use of primitive defence mechanisms, such as omnipotent denial and displacement. In this sense, it would be probably more accurate to talk about displacement children than about replacement ones" (Sabbadini 532; see also Bowlby 1998: 122 and 163-5). Indeed, the second child has the position of the bereaved's new partner who, like the unnamed protagonist of *Rebecca*, is haunted by the departed one; it is, in other words, as Dalí eloquently suggests, a displacement that never quite works, at least for the new object, all the more since there is nothing of the 'first Mrs de Winter' to be known except potentialities and projected fantasies.

Let us return finally to Anzieu's own life; first, his narrative about his own situation as replacement child:

> to me [my sister] stayed little for ever, because she died at birth. So you're right to call me an 'only son' rather than an 'only child'. In practical fact I never knew her and I lived as an only child. But in everyone's minds, this wasn't so. That lost sister, who had marked their first failure, remained for a long time in my parents' thoughts and words. I was the second one, they had to take special care of me and watch over me to protect me from the tragic fate of the elder child. I suffered the consequences of their fear of repetition. At all costs I had to survive so that my parents would be justified. But it was never enough. The least attack of indigestion or smallest draught was a threat. This put me in a difficult situation, a quite unique one. I had to replace a dead girl; yet they never let me live properly. It wasn't really a paradoxical situation; let's call it an ambiguous one.
>
> […] I couldn't go out of doors without being bundled up several times over: jumper, coat, beret, muffler. The layers of my parents' care, worries and warmth never left me, even when I lived far away from home. I carried it like a weight on my shoulders. My vitality was hidden at the core of an onion, under several skins. (APP 14-15)

He is haunted by the dead sister in the form of layers of outer clothing piled over his inner core of "vitality". That this was internalized is shown by his continuing even when he no longer lived at home – but it was internalized *externally* in behaviour and habits: literally in garments he put on.

If we assume for the moment that the parents he is referring to are René and Élise – elsewhere he describes Marguerite as just as often neglectful as over-caring – then the full cause for all this is postponed in the account to five pages later.

> The child was lightly dressed, it was cold, she went up to the fire to warm herself... and was burnt alive. It was a dreadful shock for her parents and her two sisters. So my mother was conceived as a replacement for the dead child. And since she was another girl, they gave her the same name, Marguerite. The living dead, in a way... It's no coincidence that my mother spent her life finding ways to escape from the flames of hell... It was a way of accepting her fate, a tragic fate. My mother only spoke openly of this once. But I knew it as a family legend. I think her depression goes back to this untenable position. (APP 20)

The two cases in two consecutive generations are not so much parallels, then, as causative, cumulative. Didier is the result not of one lost girl but of two. The piles of outer clothing replace the light dress worn on a cold day which produced the burnt child's skin carried thereafter, along with the borrowed name, by his mother Marguerite in her fear of "the flames of hell". Again here the family crypt, rarely spoken of, is not a closed box contained within but a series of skins worn by mother and son.

The pathology of Jeanne is inherited by Marguerite who in turn passes it from her first to her second child. What Didier chooses to do is end the chain by recognizing it:

> I might put it this way – it sounds banal, but in my case it seems true: I became a psychoanalyst to care for my mother. Not so much to care for her in reality, even though I did succeed in helping her, in the last quarter of her life, to find a relatively happy, balanced life. What I mean is, to care for my mother in myself and other people. To care, in other people, for this threatening and threatened mother... (APP 20)

We have traced the image of the common skin from its tangential to its nucleo-peripheric form and from its direct reference to the mother-child relation to its function for lovers and, beyond them, for the bereaved or betrayed. In this final image we return to its meaning for the act of reading and nurturing that is psychoanalysis. The image of the mother inside Didier Anzieu and his theory is not only held there but is also its own "threatening and threatened", consolatory and curative, but above all surviving skin.

Notes

1   See the autobiography of Amos Oz (Oz [2003] 2005), of which he said in an interview broadcast on 30 November 2005 on BBC1: "I wrote the book very much putting myself not only in the shoes but under the skin of my mother".

2   Freud's term is *'unheimlich'*, rendered into English since Strachey as 'uncanny'; but in French it is translated as *'inquiétante étrangeté'* [disturbing strangeness] and Anzieu's point here depends on the literal wording.

3   The dating of these events differs between APP and Allouch on the one hand and, on the other, the written account in Parot & Richelle, the latter running a year later (assistant to Lagache 1952-54 and in analysis with Lacan 1950-54); I have kept to the earlier dates since this accords, for instance, with Anzieu's intervention at the Rome discourse in 1953.

4   I am indebted to psychoanalyst Jane Temperley for the suggestion in a private conversation that, in discovering Lacan had analysed his mother before analysing him, Anzieu found himself for the second time in the position of the replacement child.

5   There are some similarities, especially in the key question of the 'transgenerational transmission' (whether of training or simply of analysis) between analyst and analysand, in the recent discussion in the British Psychoanalytic Association about problems concerning Donald Winnicott, Masud Khan and Wynne Godley; see especially Godley and Sandler.

6   Women were, of course, also among those who became "rapidly politicized" at this point; Anzieu notes that they were "in the majority in the Faculty of Letters"; but his psychoanalytic reading of the sexual aspects of the movement fails, probably accurately, to see the emancipation of female sexuality as a motive, except for the young men; it is surely a male

fantasy that he quotes, in a famous piece of graffiti: "The more I make revolution the more I feel like making love" (43).

7   See also APP 87: "we were a joyful bunch of babies running wild, safely grasping on to the generous body of a mother vast enough for each one to find their place and make themselves the cartographer of a particular anatomical or geographical area [or] symbolically exploring, not only the surface now but the inside of the maternal body projected into the group-object".

8   The two terms 'ego ideal' and 'ideal ego' look so similar that it is useful to distinguish them. The ego ideal (in French *Idéal du moi*) is similar to the paternal superego in that it stands as a controlling image inside the psyche, formed at the time of the breakdown of the oedipus complex; but unlike the superego, its role is positive rather than censorious. The ideal ego (in French *moi idéal*) is an earlier, narcissistic formation: it belongs to the period where the infant begins to perceive its mother as a whole object distinct from itself, and "its function is much more affective than representative" (AGI 95). Anzieu goes on: "the group illusion derives from the substitution of a common ideal ego for the individual ego ideal of each participant" (96); this substitution may often, however, be provisional or difficult.

9   For other uses of *Le Moi-peau*, see (alphabetically): Ahmed & Stacey, Benthien, Connor 2004, Deleuze & Guattari 1972, Grosz, Moorjani, Prosser, Segal 1998; Syrotinski & Maclachlan. Unfortunately the full 1995 version does not exist in English, and the translation of the 1985 version by Chris Turner in 1989 is now out of print. It is interesting to note that the term 'skin ego' – *Hautich* - was first coined, according to Claudia Benthien, by Robert Musil in his notes to *Der Mann ohne Eigenschaften* (Benthien 208).

10   Freud is referring here to the 'complete' oedipus complex, in which a child desires not only the parent of the opposite sex but also the one of the same sex (and is jealous correspondingly of both); Strachey cites in the footnote a letter to Fliess of 1 August 1899: "Bisexuality! I am sure you are

right about it. And I am accustoming myself to regarding every sexual act as an event between four individuals" (F-Ego 373). See also Butler 57-72. Recent research into the condition known as chimerism – see *The Twin inside me*, Channel 5 Monday 6 March 2006 – has shown that indeed some individuals, for instance hermaphrodites, are the literal product of four sets of DNA, variously detectable in different parts of their body.

11 For instance, when an auto-erotic impulse creates a fantasy where "I am by myself my own wife/woman [*femme*] who dreams and loves me" (ACon 143), a "sand lady" is conjured by desire on an empty beach (201) or, monstrously, "A beast, at night" takes over the protagonist's body space like an invasive version of Kafka's giant bug, but feminine and other, making him "the victim – or the accomplice – of an unknown ecstasy [*jouissance*]" (19). In another, at the end of the first five days of creation, God realizes that everything he has made so far is "in five variants, the body of a woman" and that "everything he had made was fashioned in his own image and thus he was a woman. Before he died he had just enough time to make a sensitive appendix to his creature, a companion, man, so that he could make her his god and she could pass on to him her unwanted virility" (137). In these ways, the premise of femininity as other or alien is tested in a universalizing fantasy: if "all is woman", then subjectivity must be part of it – but is the female herself and the self-feminizing impulse a deity or a beast?

12 Prokaryotic cells or organisms are those that have no nuclear membrane; eukaryotic cells or organisms "are characterised by a discrete nucleus with a membrane" (Brown 860).

13 It also corresponds to the stress laid by Freud on the discovery of the father's role, which appears in two places, *Moses and Monotheism* (F-Moses 360-1) and a footnote to the 'Rat-man' case history (F-Rat 113n): "a great advance was made in civilization when men decided to put their inferences upon a level with the testimony of their senses and to make the step from matriarchy to patriarchy". This point clashes somewhat, however, with a

theory of psychosexual development based on a fateful preference for the visible male genitals over the 'invisible' female ones.

14   This points implicitly to the epigram I have used as the epigraph to this book: the 359th of Chamfort's 1796 *Maximes et pensées* (Chamfort 127). It is cited also in *Corydon* (Gide [1911] 1993: 61) and in Sartre (SEN 430), see this book chapter 8.

15   Extraordinary as the premise of Kafka's story tends to seem, hunger artists really did exist in eighteenth- and nineteenth-century Europe – see Gooldin 27-53. In 1952, an American, the perfectly named Jack Wafer, "survived for 76 days on a diet of soda water and cigarettes", see Horsnell 2003; and of course David Blaine spent 44 days in a perspex box over the Thames, drinking only water, in September-October 2003. Media comments on Blaine were very similar in their mockery and suspicion to those described by Kafka.

16   The term 'bulimia' seems to have been quite flexible in its pre-contemporary usage: Ellmann quotes the OED as defining bulimia as "canine hunger" (68) and cites De Quincey in 1823 describing a man who ate children as bulimic; Gide uses the term, without explanation, of his friend Valery Larbaud in 1912 (GJ1 726).

17   Sources in the first three sentences, in order David Tang, cited by Johnson XIV; Woods 18; Unsigned, 'When I go home' 1; Mower 30; Attenborough 7; Kantrowitz 22; Unsigned, 'Glamorous figure' 14; Tennant, 'From shy girl' 2. In the fourth sentence, in order *The Daily Telegraph* 11; Jenkins 24; McCrum 15, Hoggart 14; Unsigned, *Vanity Fair* 186; Angelou 14, Versace 44, Hudson 12.

18   This is not to deny that Rilke himself had the fantasy, more directly represented in *Malte Laurids Brigge* and the essays on Rodin, of attaining a sort of psychological safety by creating texts that would, like a forest of sculptures, surround and protect him from everyday weakness and the

demands of other people; see the letter to Lou Andreas-Salomé of 8 August 1903 in Rilke 1952: 83-91 and Segal 1981, 83-87.

19 In the poem, like Herder, Gautier says that all male viewers will think it is Aphrodite and all female ones Cupid – just as in *Mademoiselle de Maupin*, D'Albert's horror at desiring someone he thinks is a man suggests fundamentally heterosexual laws of desire – and yet it is on the basis of these laws that the erotic is, embodied in the Hermaphrodite and in Madeleine de Maupin, what one might call post-heterosexual, depending on a proto-polymorphous doubleness.

20 Andrew Niccol wrote as well as directed *Gattaca*; Anthony Minghella adapted *The Talented Mr Ripley* from Patricia Highsmith's 1955 novel of the same name, earlier adapted to film in René Clément's *Plein Soleil* (1959); Spike Jonze shares the creative credit for *Being John Malkovich* with screenwriter Charlie Kaufman, whose more recent *Adaptation* (2002) and *Eternal Sunshine of the Spotless Mind* (2004) show similar exploration of the 'inside' of emotional heads. Incidentally, the French title of *Being John Malkovich* is *Dans la Peau de John Malkovich*.

21 Matt Damon apparently insisted that nothing Tom says should be an actual lie, so when he is accused in the Venice interrogation of killing first Freddie and then Dickie he can honestly say no.

22 See the DVD version, Kaufman 2000, and also Kaufman's original screenplay: http://www.geocities.com/Hollywood/Lot/9161/00s/BeingJohn Malkovich/beingjohnmalkovich.html; see also Litch 67-85.

23 In relation to the 'John Malkovich character, I shall call him 'John' when considering him as the fictional subject of desire, 'Malkovich' when thinking of him as the body that others enter, and 'John Malkovich' when referring to the real-life actor playing this role.

24 These quotations are all from the original screenplay, and the words are slightly different in the film, though the sexual analogy is still there.

Craig/Malkovich says to an admiring Maxine: "I've finally figured out how to hang on as long as I want. It's all a matter of making friends with the Malkovich body – rather than thinking of it as an enemy that has to be pounded into submission. I've been imagining it as a really expensive suit that I enjoy wearing".

25   I am grateful to a number of correspondents on *francofil* who answered my query in January 2006 about the term 'décoller' having the underlying meaning of 'ungluing'; in this transitive form, it dates back to 1382, but the intransitive form used by Proust and Anzieu was introduced ca 1910. Edward Forman noted: "I remember from old war movies that the speed you have to reach before taking off in a plane is referred to in English as the 'unstick speed'". The most extreme version of this unsticking is escape velocity, the speed required, in physics, to take an object out of the orbit of its source gravitational field. A composition of that name by Benjamin Wallfisch was premiered on 2 September in the 2006 BBC Proms.

26   The term 'mucous' is, as Margaret Whitford points out, never theorized (Whitford 163). As an addition to the idea of the caress it takes it vividly beyond the skin and identifies it with the feminine in a more direct way than we find elsewhere. Yet in a response to an interesting paper on applying the "morphology of the mucous" (Hilary Robinson, in Irigaray, 2002: 95) to women's painting, Irigaray rejects this extrapolation: "it is not appropriate [...] to erode skin in order to transform it into a mucous membrane. Skin must protect the internal mucous world. It is its morphology which permits the economy of mucous to exist" (101-02). The gap or interval is essential, for Irigaray: it may tend towards zero but it also maintains the balance between the two sexes.

27   For "reciprocal inclusion" see AMP 262; AEN 61, 138; ACD 226; APo 120; for "mutual inclusion" AMP 59, 122; ACD 211; not forgetting Annie Anzieu's "reciprocal incorporation" (AA 130). Didier Anzieu draws on the work of Sami-Ali, whom he calls the "father" of the skin-ego (AEN 55). In this structure, "my mother surrounds me and I surround her at the same time" (AEN 61; also APo 120; see SAE 38, 64; SAC 33-5).

28 In the screenplay his words are "Do you like me?" and he recognizes her refusal with "disappointment and despair: 'Why? Why not?'" (93); in the film the scene closes on his unanswered question.

29 Creeley himself, along with his critic Cynthia Dubin Edelberg, thought differently. He states in a 1975 interview: "'The Business' is an ironic characterization of a love imagined as a transactionary gain. [...] I was thinking of certainly then-frequent senses of exchange humanly – that kind of attitude toward marriage as being a bargain. And so the poem has that ironic tone in it" (cited Edelberg 166). She comments: "Flippantly resigned to a tenuous relationship, to getting 'The Business', the protagonist seems half-amused by his problems as his wry tone suggests" (29) and later, with more nuance, "The poems which concern women [...] are basically poignant expressions although they are superficially cynical ones [...], filled with both calculated irony and genuine desperation" (34).

Bibliography

Abraham, Nicolas and Maria Torok. [1978, 1987] 2001. *L'Écorce et le noyau*, ed Nicholas Rand. Paris: Flammarion.

Ahmed, Sara & Jackie Stacey (eds). 2001. *Thinking Through the Skin*. London and New York: Routledge.

Alderson, Andrew et al. 1997. 'A friend for life to the sick and hurting'. *The Sunday Times*. (7 September).

Allouch, Jean. [1990] 1994. *Marguerite, ou l'Aimée de Lacan*. Paris: EPEL.

Alpers, Svetlana. [1983] 1989. *The Art of Describing*. Harmondsworth: Penguin.

Alter, Jonathan. 1997. 'Diana's real legacy'. *Newsweek*. (15 September).

Amis, Martin. 1997. 'The mirror of ourselves' *Time* (15 September).

Anderson, Mark. 1992. *Kafka's Clothes*. Oxford: Clarendon.

Andrew, Geoff. 1993. 'Grand entrance'. *Time Out* (20-27 October), 24.

Angelou, Maya. 1997. Poem dated 5 September. *The Guardian*. (6 September).

Anzieu Didier. [1987] 1996a. *Les Enveloppes psychiques*. Paris: Dunod.

Anzieu, Annie. 1989. *La Femme sans qualité*. Paris: Dunod.

Anzieu, Didier & Catherine Chabert. [1961] 2003. *Les méthodes projectives*. Paris: PUF.

Anzieu, Didier [Épistémon]. 1968. *Ces Idées qui ont ébranlé la France*. Paris: Fayard.

Anzieu, Didier and Jacques-Yves Martin. [1968] 1997. *La Dynamique des groupes restreints* Paris: PUF.

Anzieu, Didier et al. 1992. *Portrait d'Anzieu avec groupe* Paris: Hommes et Perspectives.

Anzieu, Didier et al. 1993. *Les Contenants de pensée*. Paris: Dunod.

Anzieu, Didier et al. 1994c. *L'Activité de la pensée* [republished in 2000 as *Émergences et troubles de la pensée*]. Paris: Dunod.

Anzieu, Didier et al. 1999c. *Autour de l'inceste*. Paris: Collège de Psychanalyse groupale et familiale.

Anzieu, Didier. [1956] 1994. *Le Psychodrame analytique*. Paris: PUF.

Anzieu, Didier. [1959] 1998. *L'Auto-analyse de Freud et la découverte de la psychanalyse*. Paris: PUF.

Anzieu, Didier. [1975] 1995b. *Contes à rebours*. n.p.: Les Belles Lettres-Archimbaud.
Anzieu, Didier. [1975] 1999a. *Le Groupe et l'inconscient* (Paris: Dunod).
Anzieu, Didier. [1985] 1995. *Le Moi-peau*. Paris: Dunod.
Anzieu, Didier. [1998] 1999b. *Beckett*. Paris: Gallimard.
Anzieu, Didier. 'Le Moi-peau'. 1974. *Nouvelle revue de psychanalyse*, 9, 195-203.
Anzieu, Didier. 1981. *Le Corps de l'œuvre*. Paris: Gallimard.
Anzieu, Didier. 1989. *The Skin Ego*, tr Chris Turner. New Haven: Yale University Press.
Anzieu, Didier. 1990. *L'Epiderme nomade et la peau psychique*. Paris: Les Editions du Collège de psychanalyse groupale et familiale.
Anzieu, Didier. 1991. *Une Peau pour les pensées*. Paris: Apsygée.
Anzieu, Didier. 1994b. *Le Penser* . Paris: Dunod.
Anzieu, Didier. 1996b. *Créer/Détruire*. Paris: Dunod.
Anzieu, Didier. 2000. *Psychanalyser*. Paris: Dunod.
Anzieu, Marguerite. 1999. *Aimée*. Paris: atelier EPEL.
Appleyard, Bryan. 1997. 'The triumph of magic over the message'. *The Sunday Times*. (7 September).
Attenborough, Richard. 'Under the spell of a rare and illuminating personality'. *The Times*. (6 September).
Baron-Cohen, Simon. 2002. 'The extreme male brain theory of autism', *TRENDS in Cognitive Sciences*. 6:6. (June), 248-54.
Barrie, J. M. [1902]. *The Little White Bird*. London: Hodder and Stoughton.
Barrie, J. M. [1911] 1988. *Peter Pan and Wendy*. London: Pavilion.
Barthes, Roland. 1973. *Le plaisir du texte*. Paris: Seuil.
Baudelaire, Charles. [1857] 1968. *Les Fleurs du mal*, in *Œuvres complètes*, ed. Marcel Ruff. Paris: Seuil, 40-131.
Beauvoir, Simone de. 1949. *Le deuxième Sexe*, vol 1. Paris: Gallimard.
Bensoff, Harry M. 2004. 'Reception of a queer mainstream film' in Michele Aaron (ed). *New Queer Cinema*. Edinburgh: Edinburgh UP, 172-86.
Benthien, Claudia. [1999] 2002 *Skin*, tr Thomas Dunlap. New York and Chichester: Columbia UP.
Bergson, Henri. [1939] 2004. *Matière et mémoire*. Paris: PUF.
Bergson, Henri. [1940] 1981. *Le Rire*. Paris: PUF.

Bibliography

Bernasconi, Robert & Simon Critchley (eds). 1991. *Re-Reading Levinas*. London: Athlone.

Bersani, Leo. 1977. *Baudelaire and Freud*. Berkeley, Los Angeles and London: University of California Press.

Bick, Esther. 1968. 'The experience of the skin in early object-relations'. *Int. J. Psycho-Anal.* (49), 484-86.

Bignell, Jonathan. 2003. 'Where is Action Man's penis?' in Naomi Segal, Lib Taylor and Roger Cook (eds). *Indeterminate Bodies*. Basingstoke: Palgrave Macmillan, 36-47.

Bion, Wilfred. [1962] 1984. *Learning from Experience*. London: Karnac.

Birkin, Andrew. [1979] 2003. *J. M. Barrie and the Lost Boys*. New Haven and London: Yale University Press.

Bloch, Marc. 1973. *The Royal Touch* [*Les Rois thaumaturges*, 1961]. tr. J. E. Anderson. London and Montreal: Routledge & Kegan Paul.

Bloechl, Jeffrey. 2000. *The Face of the Other and the Trace of God: Essays on the Philosophy of Emmanuel Lévinas*. New York: Fordham University Press.

Blum, Virginia. 2003. *Flesh Wounds*. Berkeley: University of California Press.

Bollas, Christopher. 2000. *Hysteria*. London and New York: Routledge.

Bowlby, John. [1980] 1998. *Attachment and Loss*, vol 3: *Loss, Sadness and Depression*. London: Pimlico.

Bowlby, John. 1969, 1973 and 1980. *Attachment and Loss*. 3 vols. London: Hogarth.

Brace, Marianne. 1994. [no title, on *The Piano*]. *The Guardian*, 18 April 1994, 11.

Breton, André. [1924] 1979. '[Le premier] Manifeste du surréalisme'. *Manifestes du surréalisme*. Paris: Gallimard, 11-64.

Brown, Lesley (ed). 1993. *The New Shorter Oxford English Dictionary*. Oxford: Clarendon.

Bruzzi, Stella and Richard Cummings. 1994. Correspondence on *The Piano* in *Sight & Sound*, Feb, 72 and March, 64.

Bruzzi, Stella. 1995. 'Tempestuous petticoats: costume and desire in *The Piano*'. *Screen*. 36:3, 257-66.

Büchner, Georg. [1839] 1967. 'Lenz', in *Sämtliche Werke, Briefe und Dokumente*, vol 1, Henri Poschmann and Rosemarie Poschmann (eds). Frankfurt: Deutscher Klassiker Verlag, 223-50.

Burchill, Julie. 1997. 'She showed up the House of Windsor for what it was: a dumb, numb dinosaur'. *Evening Standard*. (3 September), 14.

Burke, Carolyn et al. 1994. *Engaging with Irigaray*. New York: Columbia.

Burston, Paul. 1997. [no title]. *Time Out*. (3-10 September), 12.

Butler, Judith. 1990. *Gender Trouble*. New York and London: Routledge.

Cairns, Lucille (ed). 2002. *Gay and Lesbian cultures in France*. New York: Peter Lang.

Campbell, Bea. 1997. *After Dark discussion*. Channel 4. (13 Sep).

Campbell-Johnston, Rachel. 1997. 'Portraits of Diana', *The Times* 5 September 1997.

Campion, Jane (dir). [1992] 1999. *The Piano* DVD. CiBy 2000.

Campion, Jane. 1993. *The Piano* screenplay. London: Bloomsbury.

Camus, Albert. 1956. *La Chute*. Paris: Gallimard.

Castañeda, Claudia. 2001. 'Robotic skin: the future of touch?', in Sara Ahmed and Jackie Stacey (eds). *Thinking Through the Skin*. London and New York: Routledge, 223-36.

Chabert, Catherine et al. 2007. *Didier Anzieu: Le Moi-peau et la psychanalyse des limites*. Paris: Éditions érès.

Chabert, Catherine. 1996. *Didier Anzieu*. Paris: PUF.

Chamfort, Sébastien-Roch Nicolas. [1796] 1923. *Maximes et pensées*. Paris: Éditions G. Crès.

Chancellor, Alexander. 1997. *The Guardian Weekend*. (6 September).

Chaney, Lisa. [2005] 2006. *Hide-and-Seek with Angels: A Life of J. M. Barrie*. London: Arrow.

Chanter, Tina (ed). 2001. *Feminist Interpretations of Lévinas*. Pennsylvania: Pennsylvania University Press.

Chartres, Rev Richard. 2007. 'Time to let her rest in peace'. *The Times*. (1 September), 5.

Clewell, Tammy. 2004. 'Mourning beyond melancholia: Freud's psychoanalysis of loss'. *Journal of the American Psychoanalytic Association*. 52:1. (March), 43-67.

Cohen, Nick et al. 1997. 'How Blair gave the monarchy a Millbank makeover', *The Observer* 7 September.
Cohen, Richard A (ed). 1986. *Face to Face with Lévinas*. Albany: State University of New York Press.
Condillac, Étienne Bonnot, Abbé de. [1754] 1984. *Traité des sensations*. Paris: Fayard.
Conley, Katharine. 1996. *Automatic Woman: The Representation of Woman in Surrealism*. Lincoln and London, University of Nebraska Press.
Connor, Steven. 2001. 'Mortification' in Sara Ahmed & Jackie Stacey (eds). *Thinking Through the Skin*. London and New York: Routledge, 36-51.
Connor, Steven. 2004. *The Book of Skin*. London: Reaktion.
Corso, Gregory. quoted at http://geocities.com/terry_young/corso.html.
Coward, Ros. 1997. [no title]. *The Guardian*. (2 September).
Critchley, Simon & Robert Bernasconi (eds.) 2002. *The Cambridge Companion to Lévinas*. Cambridge: CUP.
Dalí, Salvador. 1973. *Comment on devient Dali [sic]*, as told to André Parinaud. Paris: Robert Laffont.
Davies, Paul. 1993. 'The face and the caress' in David Michael Levin (ed). *Modernity and the Hegemony of Vision*. Berkeley, Los Angeles & London: University of California Press, 252-72.
Delay, Jean. 1957. *La Jeunesse d'André Gide*, vol 1. Paris: Gallimard.
Deleuze, Gilles and Félix Guattari. 1972. *L'Anti-Œdipe*. Paris: Minuit.
Deleuze, Gilles and Félix Guattari. 1980. *Mille plateaux*. Paris: Minuit.
Derais François and Henri Rambaud. 1952. *L'Envers du Journal de Gide*. Paris: Le nouveau portique.
Derrida, Jacques. 2000. *Le Toucher, Jean-Luc Nancy*. Paris: Galilée.
Detambel, Régine. 2007. *Petit éloge de la peau*. Paris: Gallimard.
Diderot, Denis. 'Paradoxe sur le comédien' [1773] 1951. *Œuvres*, ed André Billy. Paris: Gallimard, 1003-58.
Dieckmann, Katherine. 1992. *Interview*. 27:1 (January), 82.
Documentary on Diana's funeral. 1997. Channel 4. 3 September.
Douglas, Mary. 1966. *Purity and Danger*. London: Routledge & Kegan Paul.
Dunbar, Janet. 1970. *J. M. Barrie: the man behind the image*. London: Collins.

Dyson, Lynda. 1995. 'The return of the repressed? Whiteness, femininity and colonialism in *The Piano*'. *Screen*. 36: 3, 267-76.

Edelberg, Cynthia Dubin. 1978. *Robert Creeley's Poetry: A critical introduction*. Albuquerque: University of New Mexico Press.

Edelman, Lee. 1994. *Homographesis*. New York: Routledge.

Ellmann, Maud. 1993. *The Hunger Artists*. London: Virago.

Ernaux, Annie & Marc Marie. 2005. *L'Usage de la photo*. Paris: Gallimard.

Fiedler, Leslie. [1960], 1970. *Love and Death in the American Novel*. London: Jonathan Cape.

Field, Tiffany. [2001] 2003. *Touch*. Cambridge Mass: MIT Press.

Fisher, Seymour. [1973] 1976. *Body Consciousness*. Glasgow: Fontana/Collins.

Flaubert, Gustave. [1853] 1980. *Correspondance II*, ed. J. Bruneau. Paris: Gallimard.

Foucault, Michel. 1975. *Surveiller et punir*. Paris: Gallimard.

Fox, Robert et al. 1997. 'World leaders "saddened and devastated"'. *The Daily Telegraph*. (1 September).

Francke, Lizzie. 1993. [no title]. *Sight & Sound* (November), 224.

Franks, Alan. 1997. Interview with Madonna. *The Times*. (3 September).

Frears, Stephen (dir). 2006. *The Queen*.

Freedman, Ralph. 1996. *Life of a Poet: Rainer Maria Rilke*. New York: Farrar, Straus & Giroux.

Freud, Anna. [1967] 1998. *Selected Writings by Anna Freud*, eds Richard Ekins and Ruth Freeman. Harmondsworth: Penguin.

Freud, Sigmund. [1900] 1976. *The Interpretation of Dreams* [*Die Traumdeutung*]. tr James Strachey, *The Pelican Freud Library*, 4. Harmondsworth: Penguin.

Freud, Sigmund. [1909] 1979. 'Notes upon a case of obsessional neurosis (The "rat man")' [Bemerkungen über einen Fall von Zwangsneurose], tr James Strachey, *The Pelican Freud Library*, 9. Harmondsworth: Penguin, 33-128.

Freud, Sigmund. [1910] 1985. 'Leonardo da Vinci and a memory of his childhood' [Eine Kindheitserinnerung des Leonardo da Vinci]. tr James Strachey, *The Pelican Freud Library*, 14. Harmondsworth: Penguin, 143-231.

Bibliography

Freud, Sigmund. [1915, 1917] 1984a. 'Mourning and Melancholia' ['Trauer und Melancholie'] tr. James Strachey, *The Pelican Freud Library*, 11. Harmondsworth: Penguin, 245-68.

Freud, Sigmund. [1923] 1984b. *The Ego and the Id* [*Das Ich und das Es*], tr James Strachey, *The Pelican Freud Library*, 11. Harmondsworth: Penguin, 339-407.

Freud, Sigmund. [1933] 1973. 'Femininity' [Die Weiblichkeit], tr James Strachey, *The Pelican Freud Library* 2. Harmondsworth: Penguin, 144-69.

Freud, Sigmund. [1937-38] 1985. *Moses & Monotheism* [*Der Mann Moses und die monotheistische Religion: drei Abhandlungen*], tr James Strachey, *The Pelican Freud Library*, 13. Harmondsworth: Penguin, 237-386.

Freud, Sigmund. [1937-39] 1964. *The Standard Edition: Complete Psychological Works of Sigmund Freud XXIII*, tr/ed., James Strachey. London: Hogarth & Institute of Psycho-Analysis.

Freud, Sigmund. [1941] 1946. *Gesammelte Werke, chronologisch geordnet, 17: Schriften aus dem Nachlaß*. London: Imago.

Fried, Michael. 1967. 'Art and objecthood'. *Artforum*. 5 (June), 12-23.

Gabbard, Krin. 2004. *Black Magic: White Hollywood and African American Culture*. New Brunswick, New Jersey and London: Rutgers University Press.

Gautier, Théophile. [1835-36] 1973. *Mademoiselle de Maupin*. Paris: Gallimard.

Gautier, Théophile. [1872] 1947. *Émaux et camées*. Lille and Geneva: Droz.

Gerrard, Nicci. 1997. 'The beatification of Diana'. *The Observer*. (7 September).

Gide, André. [1891, 1952] 1986. *Les Cahiers et les poésies d'André Walter*. Paris: Gallimard.

Gide, André. [1911, 1922, 1924] 1993. *Corydon*. Paris: Gallimard.

Gide, André. 1924. *Incidences*. Paris: Gallimard.

Gide, André. 1958. *Romans; Récits et soties; Œuvres lyriques*. eds Yvonne Davet and Jean-Jacques Thierry. Paris: Gallimard.

Gide, André. 1996. *Journal 1887-1925*, ed. Éric Marty. Paris: Gallimard.

Gide, André. 1997. *Journal 1926-1950*, éd Martine Sagaert. Paris: Gallimard

Gide, André. 2001. *Souvenirs et voyages*, ed. Pierre Masson (Paris: Gallimard.

Gillett, Sue. 1995. 'Lips and fingers: Jane Campion's *The Piano*' *Screen*. 36:3, 277-87.

Gilman, Sander. 1991. *Inscribing the Other*. Lincoln & London: University of Nebraska Press.

Gilman, Sander. 1995. *Franz Kafka: The Jewish Patient*. New York: Routledge

Girard, René. 1961. *Mensonge romantique et vérité romanesque*. Paris: Grasset.

Godley, Wynne. 2001. 'Saving Masud Khan'. *London Review of Books*. 23.4 (22 Feb).

Gold, J. 1980. *An Inroduction to Behavioural Geography*. Oxford: OUP.

Goncourt, Edmond and Jules de. 1956. *Journal: Mémoires de la vie littéraire 1851-1863*, vol 1. Paris: Fasquelle et Flammarion.

Gooldin, Sigal. 2003. 'Fasting women, living skeletons and hunger artists'. *Body & Society* 9:2, 27-53.

Grant, Linda. 1997. 'Message from the Mall'. *The Guardian* (9 September).

Greer, Germaine. 2007. 'The Diana myth'. *The Sunday Times News Review* (26 August), 4-6.

Grosz, Elizabeth. 1994. *Volatile Bodies*. Bloomington and Indianapolis: Indiana University Press.

Hall, Stuart. 1997. Panellist in 'We've changed... but what will we become?'. *The Observer*. (14 September).

Hamilton, Alan. 2007. 'Service echoes with one message – time to draw a line under the tragedy'. *The Times* (1 September), 4.

Hardyment, Christina. 1988. *From Mangle to Microwave*. Cambridge: Polity.

Harrison, Kathryn. [2004] 2005. *The Mother Knot*. New York: Random House.

Heller Morton A. and William Schiff. 1991. *The Psychology of Touch*. Hillsdale, NJ, Hove & London: Lawrence Erlbaum.

Hendry, J. F. 1983. *The Sacred Threshold: A life of Rainer Maria Rilke*. Manchester: Carcanet.

Herbart, Pierre. 1952. *A la Recherche d'André Gide*. Paris: Gallimard.

Héritier-Augé, Françoise. [1985] 1989. 'Semen and blood: some ancient theories concerning their genesis and relationship', in Michel Feher (ed). *Fragments for a History of the Human Body*, Part Three. New York: Urzone, 159-75.

Highsmith, Patricia. [1955] 1999. *The Talented Mr Ripley*. London: Vintage.
Hofmannsthal, Hugo von. 1959. *Gesammelte Werk: Aufzeichnungen*. Frankfurt: Fischer.
Hoggart, Simon. 1997. 'Beauty glimpsed in a trick of the light'. *The Guardian* (6 September).
Honigsbaum, Mark. 1997. 'Holidays, dinners, presents... was this the real thing?'. *The Independent on Sunday* (7 September).
Horsnell, Michael. 2003. 'Originality of Blaine's stunt is just Wafer-thin'. *The Times* (Sep 12), 3.
Howes David (ed). 1991. *The Varieties of Sensory Experience*. Toronto, Buffalo, London: University of Toronto Press.
Howes David (ed). 2005. *Empire of the Senses: the Sensual Culture Reader*. Oxford & New York: Berg.
Hudson, Brenda. 1997. Letter to the *Bury Free Press*. (12 September).
Hugo, Victor. [1856] 1969. *Les Contemplations*, ed. Léon Cellier. Paris: Garnier.
Irigaray, Luce et al. 2002. *Dialogues. Paragraph* 25.3. (November).
Irigaray, Luce. 1974. *Speculum de l'autre femme*. Paris: Minuit.
Irigaray, Luce. 1977. *Ce Sexe qui n'en est pas un*. Paris: Minuit.
Irigaray, Luce. 1979. *Et l'une ne bouge pas sans l'autre*. Paris: Minuit.
Irigaray, Luce. 1981. *Le Corps-à-corps avec la mère*. Ottawa: Les éditions de la pleine lune.
Irigaray, Luce. 1982. *Passions élémentaires*. Paris: Minuit.
Irigaray, Luce. 1984. *Éthique de la différence sexuelle*. Paris: Minuit.
Irigaray, Luce. 1990. 'Questions à Emmanuel Lévinas'. *Critique*. 46:522 (novembre), 911-20.
Irigaray, Luce. 1997. *Être deux*. Paris: Grasset.
Jablonski, Nina G. 2006. *Skin: A natural history*. Berkeley, Los Angeles, London: University of California Press.
Jadin, Jean-Marie. 1995. *André Gide et sa perversion*. Paris: Arcanes.
James, Nick. 2000. 'My bloody Valentine'. *Sight and Sound* (February), 14-17.
Jay, Martin. 1993. *Downcast Eyes*. London, Berkeley & Los Angeles: University of California Press.

Jenkins, Simon. 1997. 'No law could have shielded her'. *The Times* (1 September).
Jenkins, Simon. 1997. 'Why do the young mourn her?'. *The Times.* (3 September).
Jobling, Ray. 1988. 'The experience of psoriasis under treatment' in Robert Anderson and Michael Bury (eds). *Living with chronic illness.* London: Unwin Hyman, 225-44.
Jobling, Ray. 1992. 'Psoriasis and its treatment in psycho-social perspective'. *Rev. Contemp. Pharmacotherapy* 3, 339-45.
Jobling, Ray. 2000. 'Psychosocial issues in dermatology' in Esther Hughes and Julie Van Onselen (eds). *Dermatology Nursing.* Edinburgh: Churchill Livingstone, 93-101.
Johnson, Boris. 1997. 'It wasn't just that she was beautiful and famous. She had this gift'. *The Daily Telegraph: Special Supplement* (1 September).
Jonze, Spike (dir). 2000. *Being John Malkovich* DVD. Universal Studios: Columbia Tristar.
Josipovici, Gabriel. 1996. *Touch.* New Haven & London: Yale University Press.
Kaës, René et al. 2000. *Les Voies de la psychè: hommage à Didier Anzieu.* Paris: Dunod.
Kafka, Franz. [1924] 2001. *Die Erzählungen.* Frankfurt: Fischer.
Kantorowitz, Barbara et al. 1997. 'The Day England Cried'. *Newsweek* (15 September).
Kapoor, Anish with Donna De Salvo. 2002. *Marsyas.* London: Tate Publishing.
Katz, Claire Elise. 2001. 'Reinhabiting the house of Ruth' in Tina Chanter (ed). *Feminist Interpretations of Emmanuel Lévinas.* Pennsylvania: Pennsylvania State University Press.
Katz, Claire Elise. 2003. *Lévinas, Judaism and the feminine.* Bloomington & Indianapolis: Indiana University Press.
Kaufman, Charlie. 2000. *Being John Malkovich.* London and New York: Faber & Faber.
Kaufman, Charlie. *Being John Malkovich,* original screenplay: http://www.geocities.com/Hollywood/Lot/9161/00s/BeingJohnMalkovich/beingjohnmalkovich.html
Keane, Molly. [1981] 2001. *Good Behaviour.* London: Virago.

Keller, James R. 2002. *Queer. (Un)friendly Film and Television*. Jefferson, North Carolina: McFarland.

Klein, Melanie. [1935] 1985a. 'A contribution to the psychogenesis of manic-depressive states', in *Love, Guilt & Reparation and Other Works 1921-1945*, ed Masud R. Khan. London: Hogarth and the Institute of Psycho-analysis, 262-89.

Klein, Melanie. [1940] 1985b. 'Mourning and its relation to manic-depressive states', in *Love, Guilt & Reparation and Other Works 1921-1945*, ed Masud R. Khan. London: Hogarth and the Institute of Psycho-analysis, 344-69.

Kleist, Heinrich von. [1810] 1990. 'Über das Marionettentheater', *Sämtliche Werke und Briefe*, vol 3, ed Klaus Müller-Salget. Frankfurt-am-Main: Deutscher Klassiker Verlag, 555-63.

Knickmeyer, Rebecca et al. 2004. 'Foetal testosterone, social relationships, and restricted interests in children', *Journal of Child Psychology and Psychiatry*. 45:0, 1-13.

Kopelson, Kevin. 1997. *The Queer Afterlife of Vaslav Nijinsky*. Stanford: Stanford University Press.

Körner, Hans. 2000. 'Die versteinerte Niobe im Marmorbild des Praxiteles: Ein Beitrag zu einer Kunstgeschichte der Oberfläche', in Susanne Peters et al (eds). *The Humanities in the New Millennium*. Tübingen and Basel: Francke, 207-35.

Kristeva, Julia. 1980. *Pouvoirs de l'horreur*. Paris: Seuil.

Lacan, Jacques. [1932] 1975. *De la psychose paranoïaque dans ses rapports avec la personnalité*. Paris: Seuil.

Lacan, Jacques. 1966. 'Jeunesse de Gide ou la lettre et le désir', in *Ecrits*. Paris: Seuil, 739-64.

Lambert, Jean. 1956. *Gide familier*. Paris: Julliard

Lane, Christopher. 1997. 'The testament of the other: Abraham and Torok's failed expiation of ghosts'. *Diacritics*. 27:4. (Winter), 3-29.

Laqueur, Thomas. 1990. *Making Sex*. Cambridge Mass: Harvard University Press.

Leconte de Lisle, Charles-Marie-René. [1884, 1886] 1977. *Œuvres*, vol 3: *Poèmes tragiques ; derniers poèmes*, ed Edgard Pich. Paris: Société d'édition « Les belles lettres »

Leconte de Lisle, Charles-Marie-René. [1889] 1976. *Œuvres*, vol 2: *Poèmes barbares*, ed Edgard Pich. Paris: Société d'édition « Les belles lettres ».

Leder, Drew. 1990. *The Absent Body*. Chicago and London: University of Chicago Press.

Leppmann, Wolfgang. 1984. tr Russell Stockman. *Rilke: A Life*. New York: Fromm.

Lessana, Marie-Magdeleine. 2000. *Entre mère et fille: un ravage*. Paris: Fayard.

Levin, David Michael (ed). 1993. *Modernity and the Hegemony of Vision*. Berkeley, Los Angeles & London: University of California Press.

Lévinas, Emmanuel. [1947, 1979] 1983. *Le temps et l'autre*. Paris: PUF.

Lévinas, Emmanuel. [1961, 1971] 2006. *Totalité et infini*. Paris: Livre de poche.

Lévinas, Emmanuel. [1974] 2004. *Autrement qu'être ou au-delà de l'essence*. Paris: Livre de poche.

Levy, Stephen. 1997. 'World Wide Wake'. *Newsweek* (15 September).

Lincoln, Sarah. 1997. [no title] *The Observer* (7 September).

Lingis, Alphonso. 1983. *Excesses*. Albany: State University of New York.

Lingis, Alphonso. 1985. *Libido: The French existential theories*. Bloomington: Indiana University Press.

Lingis, Alphonso. 1994. *Foreign Bodies*. New York and London.

Litch, Mary M. 2002. *Philosophy through Film*. New York and London: Routledge.

Longhurst, Robyn. 2001. *Bodies: Exploring Fluid Boundaries*. London & New York: Routledge.

Lyotard, Jean-François. 1974. *Économie libidinale*. Paris: Minuit.

MacDonald, Marianne. 1997. 'She let out a scream like a wild animal: "You make my life hell!"'. *The Observer*. (7 September).

Malcolm, Janet. [1983], 2004. *Psychoanalysis: The Impossible Profession* London: Granta.

Margolis, Harriet (ed). 2000. *Jane Campion's* The Piano. Cambridge: Cambridge University Press.

Marinetti, Filippo Tommaso. [1921] 1985. *L'Alcova di acciaio*. Milan: Serra e Riva.

Marinetti, Filippo Tommaso. 1912. *The Battle of Tripoli [La Bataille de Tripoli]*. Milan: Futuristo di Poesia.

Marks, Laura U. 2000. *The Skin of the Film*. Durham and London: Duke University Press.

Marks, Laura U. 2002. *Touch*. Minneapolis & London: University of Minnesota Press.

Mars-Jones, Adam. 1993. 'Poetry in motion'. *The Independent* (29 October), 26.

Mars-Jones, Adam. 1997. 'Everyone's imaginary friend' *The Observer Review* (30 November).

Martin du Gard, Roger. 1951. *Notes sur André Gide 1913-1951*. Paris: Gallimard.

Martin du Gard, Roger. 1993. *Journal* vol 2. ed. Claude Sicard. Paris: Gallimard.

Mazis, Glenn A. 1998. 'Touch and vision: rethinking with Merleau-Ponty Sartre on the caress' in John Stewart (ed). *The Debate between Sartre and Merleau-Ponty*. Evanston, Ill: Northwestern University Press, 144-53.

McCrum, Robert. 1997. 'Language of grief'. *The Observer* (7 September).

McKnight, Sam. 1997. 'The look'. *The Sunday Times Style Supplement*. (7 September).

Meltzer, Donald et al. 1975. *Explorations in Autism*. Strath Tay: Clunie.

Meltzer, Donald. 1967. *The Psychoanalytical Process*. London: Heinemann.

Merleau-Ponty, Maurice. 1945. *Phénoménologie de la perception*. Paris: Gallimard.

Merleau-Ponty, Maurice. 1964. *Le Visible et l'invisible*. Paris: Gallimard.

Meyer, Conrad Ferdinand. 1998. *Gedichte*, ed Rüdiger Görner. Frankfurt: Insel.

Meyers, Carol. 1988. *Discovering Eve: Ancient Israelite women in context*. New York and Oxford: Oxford University Press.

Miller, Jacques-Alain. 1994. 'Extimité'. tr Françoise Massardier-Kenney. In Mark Bracher et al. (eds). *Lacanian Theory of Discourse*. New York and London: New York University Press, 74-87.

Miller, William Ian. 1997. *The Anatomy of Disgust*. Cambridge Mass and London: Harvard University Press, 1997

Millot, Catherine. 1996. *Gide Genet Mishima*. Paris: Gallimard.

Minghella, Anthony (dir). 1999. *The Talented Mr Ripley* DVD. Paramount & Miramax.
Minghella, Anthony. 2000. *The Talented Mr Ripley* screenplay. London: Methuen.
Monckton, Rosa. 1997. 'My friend Diana'. *The Guardian*. (8 September).
Monckton, Rosa. 1997. 'She found a sort of peace in life', *The Daily Telegraph*. (1 September).
Montagu, Ashley. [1971] 1986. *Touching: The human significance of the skin*. New York: Harper & Row.
Moore, Suzanne. 1997. 'Diana: unique, complex, extraordinary, irreplaceable', *The Independent on Sunday*. (7 September).
Moorjani, Angela. 2000. *Beyond Fetishism and Other Excursions in Psychopragmatics*. Basingstoke and London: Palgrave.
Morris, Jan. 1997. 'The naughty girl next door'. *Time*. (15 September).
Morton, Andrew. [1992] 1997. *Diana: Her True Story – In Her Own Words*. St Ives: Michael O'Mara.
Mower, Sarah. 1997. 'State of grace', *The Times Magazine*. (6 September).
Moyaert, Paul. 2000. 'The Phenomenology of Eros: a reading of *Totality and Infinity*, IV.B', in Jeffrey Bloechl (ed). *The Face of the Other and the Trace of God*. New York: Fordham University Press, 30-42.
Musset, Alfred de. [1834] 1976. *Lorenzaccio*. Paris: Bordas.
Musset, Alfred de. [1836] 1973. *La Confession d'un enfant du siècle*. Paris: Gallimard.
Nancy, Jean-Luc. [2000] 2006. *Corpus*. Paris: Métailié.
Nancy, Jean-Luc. 2003. *Noli me tangere*. Paris: Bayard.
Niccol, Andrew (dir). 1999. *Gattaca* DVD. London: Columbia Tristar.
Nietzsche, Friedrich [1887] 1971. *Morgenröthe*, no 48, in *Werke: Kritische Gesamtausgabe*, Part V, vol 1, eds Giorgio Colli and Mazzino Montinari. Berlin and New York: Walter de Gruyter
Nobécourt, Lorette. 1994. *La Démangeaison* Paris: Les Belles Lettres.
Nolan, Christopher, (dir). 2002. *Memento* DVD. London: Pathé.
Nolan, Christopher. 2001. *Memento* and *Following*. London: Faber and Faber.
Olivier, Edith. [1927] 1981. *The Love Child*. London: Virago.

Oz, Amos. [2003] 2005. *A Tale of Love and Darkness*, tr. Nicholas de Lange. London: Vintage.

Parot, Françoise & Marc Richelle (eds). 1992 *Psychologues de langue française*. Paris: PUF.

Parry, Idris (tr and ed). 1994. *Essays on Dolls*. London and New York: Penguin.

Paska, Roman. 1989 'The inanimate incarnate', in Michel Feher (ed). *Fragments for a History of the Human Body*, Part One. New York: Urzone, 411-30.

Peters, H. F. [1960] 1963. *Rainer Maria Rilke: Masks and the Man*. New York, Toronto, London: McGraw-Hill.

Pierre-Quint, Léon. 1952. *André Gide: L'homme, sa vie, son œuvre*. Paris: Stock.

Potts, Alex. 2000. *The Sculptural Imagination*. New Haven and London: Yale University Press.

Prater, Donald. 1986. *A Ringing Glass: The Life of Rainer Maria Rilke*. Oxford: Clarendon.

Prosser, Jay. 1998. *Second Skins*. New York: Columbia University Press.

Proust, Marcel. [1918] 1954a. *A l'ombre des jeunes filles en fleurs. A la recherche du temps perdu*, vol 1, ed. Pierre Clarac and André Ferré. Paris: Gallimard, 431-990.

Proust, Marcel. [1925] 1954b. *La Fugitive. A la recherche du temps perdu*, vol 3, ed. Pierre Clarac and André Ferré. Paris: Gallimard, 419-688.

Pryor, Ian. 1993. Interview with Jane Campion. *Onfilm*. 10 (9 Oct), 25.

Pullinger, Kate. 1995. *The Piano: A Novel*. Hyperion: Boston.

Purves, Libby. 1997. 'A simple heart in a heartless world'. *The Times*, (1 September).

Ramachandran Vilayanur. 2003. *The Emerging Mind*. London: Profile.

Ramachandran, V. S. and Sandra Blakeslee [1998] 1999. *Phantoms in the Brain*. London: Fourth Estate.

Redburn, Chris. 1997. Editorial, *Time*. (15 September).

Richards, Peter. 1977. *The Medieval Leper and his Northern Heirs*. Woodbridge and Rochester: D. S. Brewer.

Rilke, Rainer Maria. [1903, 1907] 1930a. 'Auguste Rodin', *Gesammelte Werke*, Band IV: *Schriften in Prosa: Erster Teil*. Leipzig: Insel, 299-418.

Rilke, Rainer Maria. [1907, 1908] 1974. *Neue Gedichte* and *Der Neuen Gedichte anderer Teil*. Frankfurt: Insel.
Rilke, Rainer Maria. [1910] 1975. *Die Aufzeichnungen des Malte Laurids Brigge*. Frankfurt: Suhrkamp.
Rilke, Rainer Maria. [1913-14] 1930b. 'Puppen', in *Gesammelte Werke*, Band IV: *Schriften in Prosa: Erster Teil*. Leipzig: Insel, 265-77.
Rilke, Rainer Maria. 1952. *Briefwechsel mit Lou Andreas-Salomé*. Zurich: Niehans.
Rodaway, Paul. 1997. *Sensuous Geographies*. London & New York: Routledge.
Roudinesco, Élisabeth. [1986] 1994. *Histoire de la psychanalyse en France*, 2 vols. Paris: Fayard.
Roudinesco, Élisabeth. [1994] 1997. *Jacques Lacan*, tr Barbara Bray. Cambridge: Polity.
Rowe, 'Paula' with Beverly Kemp. 2006. 'My first wife would do my hair in pigtails', in *Now*. (8 February), 33.
Rudd, Les. 1997. Cited in 'The day I met Diana'. *The Observer*. (7 September).
Ruddick, Sara. [1989] 1990. *Maternal Thinking*. London: The Women's Press.
Sabbadini, Andrea. 1988. 'The replacement child', in *Contemporary Psychoanalysis*. 24, 528-47.
Sacks, Oliver. [1984] 1991. *A Leg to Stand on*. London: Picador.
Sacks, Oliver. [1985] 1986. *The Man who mistook his wife for a hat*. London: Picador.
Saint-Exupéry, Antoine de. 1931. *Vol de nuit*. Paris: Gallimard.
Sami-Ali, Mahmoud. [1977] 1998. *Corps réel, corps imaginaire*. Paris: Dunod.
Sami-Ali, Mahmoud. 1974. *L'Espace imaginaire*. Paris: Gallimard.
Sandford, Stella. 2002. 'Lévinas, feminism and the feminine', in Simon Critchley and Robert Bernasconi (eds), *The Cambridge Companion to Lévinas*. Cambridge: Cambridge University Press, 139-60.
Sandler, Anne-Marie. 2004. 'Institutional responses to boundary violations', *International Journal of Psychoanalysis*. 85. 1 (February), 27-43.
Sartre, Jean-Paul. 1939. *Le Mur*. Paris: Gallimard.
Sartre, Jean-Paul. 1943. *L'Être et le néant*. Paris: Gallimard.
Sartre, Jean-Paul. 1951. *Le Diable et le bon Dieu*. Paris: Gallimard.

Sartre, Jean-Paul. 1971. *L'Idiot de la famille: Gustave Flaubert de 1821 à 1857* vol 2. Paris: Gallimard.

Schilder, Paul. [1950] 1978. *The Image and Appearance of the Human Body*. New York: International Universities Press.

Schnack, Ingeborg. ed. 1973. *Rainer Maria Rilke: Leben und Werk im Bild*. Frankfurt: Insel.

Sedgwick, Eve Kosofsky. 1985. *Between Men*. New York: Columbia University Press.

Sedgwick, Eve Kosofsky. 2003. *Touching Feeling*. Durham & London: Duke University Press.

Segal, Naomi. 1981. *The Banal Object*. London: Institute of Germanic Studies.

Segal, Naomi. 1988. *Narcissus and Echo: women in the French récit*. Manchester: Manchester University Press.

Segal, Naomi. 1992. *The Adulteress's Child: authorship and desire in the nineteenth-century novel*. Cambridge: Polity.

Segal, Naomi. 1997. 'The Fatal Attraction of *The Piano*', in Nicholas White and Naomi Segal (eds), *Scarlet Letters*. London: Macmillan, 199-211.

Segal, Naomi. 1998a. *André Gide: Pederasty and Pedagogy*. Oxford: Oxford University Press.

Segal, Naomi. 1998b. 'The common touch', in ed. Mandy Merck, *After Diana*. London and New York: Verso, 131-45.

Segal, Naomi. 2001a. '"L'échange de deux fantaisies et le contact de deux épidermes": skin and desire', in Michael Syrotinski and Ian Maclachlan (eds) *Sensual Reading*. Lewisburg and London: Associated University Presses, 17-38.

Segal, Naomi. 2001b. 'André Gide et les garçons perdus'. *Bulletin des Amis d'André Gide* 29:131-132 (juillet-octobre), 355-77.

Segal, Naomi. 2002a. 'Gide in Egypt 1939' in Charles Burdett and Derek Duncan (eds), *Cultural Encounters: European Travel Writing in the 1930s*. New York & Oxford: Berghahn, 143-55.

Segal, Naomi. 2002b. 'The indiscreet charm of the hysteric'. *Psychoanalysis and History* 4:2, 235-44.

Segal, Naomi (ed). 2000. *Le Désir à l'œuvre: André Gide à Cambridge 1918, 1998*. Amsterdam & Atlanta: Rodopi.

Serres, Michel. 1985. *Les cinq Sens*. Paris: Grasset.

Shildrick, Margrit. 1997. *Leaky Bodies and Boundaries: Feminism, Postmodernism and (Bio)Ethics*. London, New York: Routledge.

Showalter, Elaine. 1997. 'Storming the wintry palace'. *The Guardian* (6 September).

Smith, Marquard and Joanne Morra. 2006. *The Prosthetic Impulse*. Cambridge Mass and London: MIT Press.

Sontag, Susan. [1977, 1988], 1991. *Illness as metaphor; AIDS and its metaphors*. Harmondsworth: Penguin.

Street, Sarah. 2001. *Costume and cinema: dress codes in popular film*. London and New York: Wallflower.

Strong, Roy. 1997. 'Icon for our times'. *The Times Magazine* (6 September).

Syrotinski, Michael & Ian Maclachlan (eds). 2001. *Sensual Reading*. Lewisburg and London: Bucknell University Press & Associated University Presses.

Tennant, Laura. 1997. 'From shy girl to a woman at ease with herself'. *The Independent on Sunday Real Life* (7 September).

Tennant, Laura. 1997. 'The incomparable queen of glamour'. *The Independent on Sunday: Real Life*. (7 September).

Thatcher, Margaret. 1997. cited in *The Daily Telegraph*. (1 September).

Tolstoy, Leo. [1869] 2005. *War and Peace*, tr Anthony Briggs. Harmondsworth: Penguin.

Turkle, Sherry. [1978] 1992. *Psychoanalytic Politics: Jacques Lacan and Freud's French Revolution*. New York and London: Free Association.

Tustin, Frances. [1981] 2003. *Autistic States in Children*. Hove and New York: Brunner-Routledge.

Tyler, Imogen. 2001. 'Skin-tight: celebrity, pregnancy and subjectivity', in Sara Ahmed and Jackie Stacey (eds). *Thinking Through the Skin*. London and New York: Routledge, 69-83.

Unsigned caption. 1997. *Hello*. (15 September).

Unsigned editorial. 1997. 'Media, monarchy and the Earl', *The Guardian*. (8 September).

Unsigned Editorial. 1997. 'Who does not have a share of this?' *The Independent on Sunday*. (7 September).

Unsigned editorial. 1997. *The Times*. (1 September).

Unsigned. 1997. '"When I go home and turn my light off at night, I know I did my best"'. *The Times Magazine*. (6 September).
Unsigned. 1997. 'Glamorous figure who turned heads towards good causes'. *The Times.* (1 September).
Unsigned. 1997. 'Nation learns to grieve in a week of wonders'. *The Guardian.* (6 September).
Unsigned. 1997. *Vanity Fair* (December).
Unsigned. 2007. 'Empty seats at service show scars of Diana's life have not all healed'. *The Times.* (31 August), 7.
Valéry, Paul. 1960. 'L'idée fixe', *Œuvres complètes.* Paris: Gallimard.
Van Rysselberghe, Maria. 1973, 1974, 1975 & 1977. *Les Cahiers de la Petite Dame. Cahiers André Gide* vols 4-7. Paris: Gallimard.
Vasseleu, Cathryn. 1998. *Textures of Light.* London and New York: Routledge.
Weiss, Gail. 1999. *Body images.* New York and London: Routledge.
White, Kate (ed). 2004. *Touch.* London: Karnac.
White, Nicholas and Naomi Segal (eds). 1997. *Scarlet Letters: Fictions of Adultery from Antiquity to the 1990s.* Basingstoke: Macmillan.
Whitford, Margaret. 1991. *Luce Irigaray: Philosophy in the Feminine.* London: Routledge.
Williams, Michael. 2004. '*Plein Soleil* and *The Talented Mr Ripley*: Sun, Stars and Highsmith's Queer Periphery'. *Journal of Romance Studies.* 4:1, 47-62.
Williamson, Judith. 1997. 'A glimpse of the void', *The Guardian Weekend.* (13 September).
Wilson, A. N. 1997. 'This cult of Diana makes me shiver'. *The Evening Standard.* (10 September).
Winnicott, D. W. 1989. *Psycho-analytic Explorations,* eds Clare Winnicott, Ray Shepherd and Madeleine Davis. London: Karnac.
Winnicott, Donald. [1952] 1984. 'Anxiety associated with insecurity', *Through Paediatrics to Psychoanalysis.* London: Karnac, 97-100.
Winnicott, Donald. [1958] 1990a. 'The capacity to be alone', *The Maturational Processes and the Facilitating Environment.* London: Karnac, 29-36.

Winnicott, Donald. [1965] 1990b. *The Maturational Processes and the Facilitating Environment*. London: Karnac.
Winnicott, Donald. [1971] 1985. *Playing and Reality*. Harmondsworth: Penguin.
Wood, Aylish. 2002. *Technoscience in contemporary American Film*. Manchester: Manchester University Press.
Woods, Vicki. 1997. 'The girl couldn't help it'. *The Guardian* (1 September).
Woolf, Virginia. [1929] 1977. *A Room of One's Own*. London: HarperCollins.
Wright Wexman, Virginia (ed). 1999. *Jane Campion Interviews*. Jackson: University of Mississippi Press.
Yeats, William Butler. 1990. *W. B. Yeats: The Poems*, ed Daniel Albright. London: Dent.
Young, Hugo. 1997. 'Not one of us at all'. *The Guardian* (2 September).
Young, Iris Marion. 1977. *Throwing like a Girl*. Bloomington: Indiana University Press.
Young-Bruehl, Elisabeth. 1988. *Anna Freud*. London: Macmillan.

Index

'A une robe rose': 122
'Aaron Green': 55
'Ada': 201-07, 211, 228, 247
'Agatha': 222-23, 225
'Aimée': 8, 18
'Albertine': 227
'Alissa Bucolin': 96, 98
'Amédée Fleurissoire': 83, 85
'Anton Freeman': 147-49, 151
'Baines': 202-07
'Barbie': 133
'Bergotte': 173, 178
'Bernard Profitendieu': 93
'Binker': 222, 223
'Bonnie': 210
'Buffalo Bill': 159
'Caesar': 146
'Candle in the wind': 114, 115
'Captain Hook': 174
'Cesare': 134
'Charles Bovary': 175
'Charlus': 86, 165
'Christof': 208, 209, 210
'Clamence': 114
'Clarissa': 222-23, 225
'Clyde': 210
'Contralto': 123-24, 125, 136
'Coppelius': 134
'Coquetterie posthume': 122, 125
'Craig': 161-68, 171, 246
'D'Albert': 124, 245

'Daniel' (Eugène Rouart): 80-81
'Das Rosen-innere': 126
'Der Ball': 126, 174
'Der römische Brunnen': 120
'Der Sandmann': 130
'Dickie Greenleaf': 145, 153-60, 245
'Director Josef': 147, 151
'Dr Caligari': 134
'Dr Lamar': 147, 152-53
'Dr Lester': 162, 163, 164, 168
'Dumbo': 172
'Édouard': 93
'Ein Hungerkünstler': 92
'Élévation': 171-72, 174, 177
'Émile X': 89-90
'Emily': 162, 164, 168, 171
'Emma Bovary': 13, 114, 144, 175
'Emmanuèle': 97, 98, 214
'Epistémon': 23, 25, 26,
'Et nunc manet in te': 215
'Eugene Morrow': 145, 146-53, 160
'Flora': 202, 203, 205-06, 222
'Francis': 134
'Freddie': 154, 155, 157, 245
'German': 151
'Gertrude': 134
'GI Jane': 163
'Gilbert Osmond': 165
'Gregor Samsa': 152

'Gribouille': 90
'Hélène': 110
'Herbert Greenleaf': 153, 157
'Impéria': 122
'Ingeborg': 128
'Inter urinam et faeces nascimur': 57
'Intimité': 169
'Irene': 146, 147, 151,
'Jerome Morrow': 145, 146-53
'Jérôme Palissier': 95, 96, 98
'John Malkovich': 145, 161-68, 171, 245, 246
'Julien Sorel': 13,
'Kaled': 124
'L'art': 123
'L'écorce et le noyau': 219-20
'L'épiderme nomade': 58-60, 159, 228-29
'La beauté': 177
'La psychanalyse encore': 27
'La Rose-thé': 122
'La structure nécessairement narcissique de l'œuvre': 42
'Lady Grizel': 143
'Lafcadio': 85-86, 89, 160
'Lara': 124
'Le poème de la femme': 122
'Le rêve d'un curieux': 172
'Le sommeil du condor': 176-77
'Le vierge, le vivace et le bel aujourd'hui': 122
'Le voyage': 174
'Lenz': 120-21
'Leonard': 227
'Léonore': 65-67
'Lorenzaccio': 144-45, 151, 159, 170
'Lotte': 161-68, 171
'Lucile Bucolin': 95-96
'Madeleine de Maupin': 125, 145, 245
'Malte Laurids Brigge': 128
'Marcel': 227
'Marceline': 90
'Marge': 153-57
'Marlon': 209
'Maxine': 161-68, 171, 246
'Mazetto': 124
'Meredith Logue': 156, 157, 160
'Meryl': 209, 210
'Michel': 90
'Michelangelo und seine Statuen': 120
'Mourning and melancholia': 217-20
'Mrs Brock': 143
'Mrs Ramsay': 112
'Nathanael': 134
'O strophe du poète': 213-14
'Olimpia': 134
'Orpheus. Eurydike. Hermes': 215-16
'Othello': 114
'Palatine': 37-38, 235-36
'Peter Pan': 160, 174, 232, 234
'Peter Smith-Kingsley': 154, 157-59
'Pinocchio': 134
'Protos': 89

Index

'Puppen': 131-33
'Rebecca': 237
'Rome discourse': 19, 241
'Romeo und Julia auf dem Dorfe': 134
'Rosalind': 124
'Rosette': 124
'Silvana': 156, 158
'Sophie': 128, 233
'Stewart': 204, 205-06, 228, 247
'Superman': 174
'Sylvia': 210
'The Business': 211
'The Rat-man': 243
'Tom Ripley': 153-59, 165, 245
'Truman Burbank': 208-10, 211, 221
'Über das Marionettentheater': 128-29
'Victor': 3
'Vincent Freeman': 146-53, 159, 178, 210
'Wendy': 231-32
'Wunsch, Indianer zu werden': 173, 179
'Zerlina': 124

Abelard: 161
Abraham, Karl: 217
Abraham, Nicolas: 46, 217, 218-21, 223, 229, 236
Acteon: 180, 186
Adaptation: 245
Ahmed, Sara: 242
Aimée: 13,

Al Fayed, Mohamed: 116, 117
Alderson, Andrew: 108
Allégret, Marc: 134
Allouch, Jean: 8, 10, 13, 241
Alter, Jonathan: 108
Amis, Martin: 102
Amiyna: 109
Anality: 42, 70,
Anderson, Mark: 92
Andre, Carl: 138, 139
Andreas-Salomé, Lou: 232, 244
Angelou, Maya: 103, 244
Angels: 102, 115, 129-30, 132, 221-22, 223
Anne, Princess Royal: 113
Anorexia: 91-93, 96-98, 101-02, 104, 112,
Anzieu, Annie (Péghaire): 17, 18, 19, 56, 60-63, 72, 75, 103-04, 190-91, 196, 202, 246
Anzieu, Christine: 19
Anzieu, Didier: 4-6, 7-77, 83-84, 86, 90, 91, 93, 102, 103, 135, 141, 144, 159, 173, 174, 175, 178, 180, 189-200, 201, 219, 228, 229, 235-36, 237-39, 241, 246
Anzieu and May '68: 22-26
Anzieu in relation to Sartre, Merleau-Ponty, Lévinas and Irigaray: 197-200
Anzieu on love: 192-95
Anzieu on the caress and sexual practice: 189-92

Anzieu on the common skin: 195-97, 239
Anzieu's early years: 7-18
Anzieu's fiction: 56-60, 77
Anzieu's relation to Lacan: 18-21, 38, 241
Anzieu's subsequent career: 26-30
Anzieu's theory of creation: 38-43, 56, 67-72, 75, 173-74
Anzieu's theory of groups: 33-37, 56, 63-67, 75, 242
Anzieu's theory of psychodrama: 31-33;
Anzieu's theory of the skin-ego: 44-49, 56
Anzieu's theory of thought: 49-52, 67
Anzieu, Marguerite (Pantaine): 7, 8, 10-13, 15, 17, 18, 19, 28, 29-30, 38, 52, 72, 77, 133, 194, 231, 233, 234, 237-39, 241
Anzieu, Pascal: 19
Anzieu, René: 9, 12, 15-17, 29-30, 52, 238
Anzieu's stillborn sister: 11, 14, 37-38, 237-38
APF (Association Psychanalytique de France): 22
Aphrodite: 124, 245
Apollo: 141
Appleyard, Bryan: 104
Archer, Lord Jeffrey: 103

Aristotle: 2, 70, 189,
As You Like It: 124
Attenborough, Richard: 244
Aunts: 86, 95-99
Author/writer: 39, 41, 161, 175,
Autism: 5, 135, 179, 187, 187, 188, 202, 207,
Baby: 4, 8, 11, 33, 36, 45-47, 51, 59, 70, 191,
Backcloth/background: 48, 182, 191, 192, 199-200,
Baker, Chet: 156
Balmond, Cecil: 139-40
Baron-Cohen, Simon: 135
Barrie, David: 231-32
Barrie, J. M.: 64, 231-32, 234
Barrie, Jane Ann: 231
Barthes, Roland: 119
Bashir, Martin: 101, 111,
Baudelaire, Charles: 3, 171-72, 174-75, 176, 177,
Bean, Orson: 165
Beauty: 58, 69, 89, 92, 102, 104, 105, 106, 110, 113, 115, 121, 123-24, 125, 126, 129, 153, 160, 165, 198, 230,
Beauvoir, Simone de: 185
Beckett, Samuel: 28-29, 39, 41, 43, 76,
Beckett, Suzanne: 28
Beckett: 28-29, 38, 43, 192, 194,
Being John Malkovich: 145, 160-68, 171, 245
Bellmer, Hans: 130
Bensoff, Harry M.: 156

Index

Benthien, Claudia: 242
Bereavement/ mourning: 16, 41, 50, 115, 116, 133, 175, 212, 215, 217-20, 221, 225, 229, 234, 236, 239
Bergson, Henri: 129
Bernard, Claude: 17, 93,
Bersani, Leo: 172, 177-78,
Betrayal: 212, 213, 221, 227, 239
Bick, Esther: 46, 47, 50, 99, 135
Bignell, Jonathan: 133-34
Bion, Wilfred: 33, 41, 43, 46, 50, 52, 74-75
Bioy Casares, Adolfo: 199
Birkin, Andrew: 231-32
Birth: 57, 61, 65, 68, 70, 73-74, 198
Bisexuality: 58, 136, 242-43
Blaine, David: 244
Blair, Tony: 117
Bloch, Marc: 109
Blood: 36, 62, 90, 106, 146, 148, 151, 225
Body: 4, 6, 31, 35, 39, 42, 46, 50, 56, 69, 72, 76, 85, 86, 88, 91, 94, 98, 114, 115, 123, 126, 136, 137, 145, 149, 152, 153, 160, 166, 169, 171, 175, 178, 183, 185, 189, 199, 203, 226,
Bollas, Christopher: 103, 190
Bonnard, Pierre: 124
Bonnaterre, Pierre-Joseph: 3
Borges, Jorge Luis: 28, 75, 196
Bourgeois, Louise: 137
Bowlby, John: 45, 231, 236

Brace, Marianne: 202
Brazelton, T. Berry: 196
Breastfeeding: 46, 51, 61, 191-92
Breton, André: 139
Breuil, Charles: 27
Brown, Lesley: 243
Bruzzi, Stella: 206
Büchner, Georg: 120-21, 122, 125
Bulimia: 91, 101-03, 105, 107, 110-12, 114, 244
Buonarroti, Michelangelo: 120
Burchill, Julie: 112
Burrell, Paul: 118
Burston, Paul: 106
Butler, Judith: 44, 243
Cain: 155
Camilla, Duchess of Cornwall: 111, 114, 117-118
Campbell, Bea: 107, 109
Campbell-Johnston, Rachel: 107
Campion, Jane: 201, 204, 206, 207
Camus, Albert: 114
Canova, Antonio: 136, 138
Capgras' syndrome: 226-27
Caress: 110, 120, 125, 130, 158, 170, 171, 180-90, 197-99, 201, 203-08, 211, 227, 228, 233, 246
Castañeda, Claudia: 208
Castration: 56, 72, 173, 174, 178
CEFFRAP: 20, 34
Cells: 137
Chabert, Catherine: 20, 32

Chamfort, Sébastien-Roch Nicolas: 244
Chancellor, Alexander: 105
Chanter, Tina: 185
Chapman, Jan: 206
Charles and Camilla: Whatever Love Means: 117
Charles and Diana: The Wedding: 118
Charles, Prince of Wales: 107, 111, 112, 113, 115, 117-18
Chartres, Rev Richard: 118
Chekhov, Anton: 165
Chimera/chimerism: 123, 243
Ciccone, Albert: 195
Cinderella: 124, 149, 154, 164
Circuit: 101-03, 105-07, 108, 111-12, 116, 119, 128, 138, 198, 209, 228
Classen, Constance: 1, 3
Clément, René: 154, 156, 245
Clewell, Tammy: 217, 229
Clifford, Max: 115
Clothes: 48, 58-59, 69, 80, 83, 86-89, 112, 122, 124, 125, 128, 132, 133, 143, 145, 158, 160, 170, 204, 205, 228, 229, 230, 233-34, 237, 238, 246
Cocteau, Jean: 134, 143
Coleridge, Samuel Taylor: 176
Colet, Louise: 143
Collodi, Carlo: 134
Common sense: 6, 48
Common skin/*peau commune*: 31, 47, 57, 75, 76, 109, 128, 137, 145, 160, 192, 195-97, 198-99, 201, 229-30, 238
Condillac, Étienne Bonnot, Abbé de: 134
Conley, Katharine: 13
Connor, Steven: 84, 107, 242
Consensuality: 5-6, 47, 48, 52, 73, 109, 119, 128, 179, 183, 184, 187, 192, 200, 204, 205, 207, 228
Constant, Benjamin: 218
Containment: 4, 31, 33, 35, 37, 46, 47, 50, 56, 73-76, 94, 98, 99, 102, 123-24, 145, 163, 182, 199, 227, 228
Contes à rebours: 28, 57-60
Coppélia: 130
Corps étranger: 170
Corso, Gregory: 139
Corydon: 244
Cotard's syndrome: 227
Counter-transference: 33, 35, 55
Couple: 6, 10, 28, 34, 55, 66, 67, 74, 192-94, 199, 239
Coward, Rosalind: 105
Crawford, Cindy: 115
Craxton, Kim: 115
Creativity/creation: 39-43, 51, 56, 67-72, 75, 77, 122, 132, 134, 144, 161-62, 173-74, 229, 244-45
Creeley, Robert: 211, 247
Créer/Détruire: 32, 42, 43,
Cronenberg, David: 169
Crustacean: 47, 89

CSI: 148
Cupid: 124, 245
Cusack, John: 161, 165
Cuverville: 98,
Dalí, Salvador: 230-31, 234, 237
Damon, Matt: 145, 153, 160, 245
Darwin, Charles: 176
Davies, Paul: 185
De Quincey, Thomas: 244
Dead Ringers: 169
Death: 7-8, 10, 13, 26, 29, 39, 41, 58, 92, 98, 112, 113, 115-16, 118, 122, 131, 154, 156, 157, 159-60, 168, 172, 186-87, 212, 214-16, 218, 230, 231
Delay, Jean: 90
Deleuze, Gilles: 179, 208, 242
Denver, John: 115
Der Himmel über Berlin: 222
Der Mann ohne Eigenschaften: 242
Derais François: 96
Derrida, Jacques: 185
Desire: 6, 59, 60, 66, 77, 83, 90, 91, 92-93, 94, 98, 99, 100, 119, 120, 124, 125, 128, 131, 145, 153, 155, 158-59, 160, 162, 166-67, 168, 169, 171, 174, 175, 177, 179-200, 205, 206, 211, 212, 222-23, 227, 228, 245
Detambel, Régine: 84, 208
Diana and the Camera: 118
Diana, Princess of Wales: 5, 32, 79, 101-18, 175, 209, 221, 228

Diaz, Cameron: 162
Diderot, Denis: 129
Die Aufzeichnungen des Malte Laurids Brigge: 128, 233-34, 244
Die Verwirrungen des Zöglings Törleβ: 128
Die Wahlverwandtschaften: 160-61
Dinggedichte: 120, 122, 126-28
Dirt: 84, 94, 96, 148,
Dolls: 62, 128-34, 161, 167, 232-33, 234
Dolto, Françoise: 21
Douglas, Lord Alfred: 152
Douglas, Mary: 94
Duino Elegies: 102, 129-30, 133, 233
Dunbar, Janet: 231
Dunod: 27
Dyson, Lynda: 204
École normale supérieure: 17
Edelberg, Cynthia Dubin: 247
Edelman, Lee: 86,
EFP (École Freudienne de Paris): 22
Ego ideal: 37, 64, 242
Ego: 41, 47, 50, 52, 101, 192, 217, 219
Eliot, T. S.: 221-22
Elizabeth the Queen Mother: 114
Ellmann, Maud: 91, 244
Émaux et camées: 121-24

Emptiness/emptying: 32, 43, 81-83, 90-93, 94, 99, 101, 113, 139, 223
Erasmus: 26
Eternal Sunshine of the Spotless Mind: 245
Eurydice: 215-16, 227
Eve: 66, 127
Existenz: 164
Face: 28, 80, 85, 86, 95, 98, 99, 103, 107, 111, 121, 126, 134, 145, 147, 149, 152, 153, 155, 165, 185, 186-87
Family: 10, 13-16, 30, 34, 38, 49, 59, 68, 77, 84, 112, 113, 117, 185, 194, 219, 230, 236, 238
Father: 37, 40, 42, 63, 97, 146-47, 157, 164, 168, 213
Faure, Edgar: 26
Favez, Georges: 19, 38,
Favez-Boutonier, Juliette: 20
Fayed, Dodi: 115, 116
Fechner, Gustav: 93
Fecundity: 61, 64, 67, 72, 187
Feedback/homeostatic loop: 46-47, 74, 192, 196
Feminine/femininity: 29, 36, 37, 39, 57, 61, 62, 63, 68, 69, 72, 74, 79, 86, 90, 94, 95, 99, 100, 102, 106-08, 112, 132-33, 134, 136, 150, 161, 162, 164, 165, 173, 180, 185-86, 190, 191, 202, 234, 235, 243, 246
Feminism: 56, 107
Ferenczi, Sándor: 55, 218,

Ferguson, Sarah: 113
Fetish/fetishism: 64, 119, 135, 158
Fiedler, Leslie: 86
Field, Tiffany: 4, 45
Filloux, Jean-Claude: 22
Fire/flames: 7, 9, 12, 49, 132-33, 150, 209, 238
Fisher, Seymour: 169
Flaubert, Gustave: 42, 143-44, 161, 175, 222
Fliess, Wilhelm: 242
Fluids/fluidity: 56, 83, 90, 91, 93-94, 96, 98, 100, 105, 106, 112, 115, 120, 132, 135, 139, 146, 187, 190-91, 206
Food: 15, 62, 111-12, 223
Formal signifier/ *signifiant formel*: 31-32, 76, 79, 169, 212
Forman, Edward: 246
Foucault, Michel: 107-08, 109,
Fox, Robert: 103
Franks, Alan: 110
Frears, Stephen: 117
Free association: 24, 57
Freedman, Ralph: 128, 233,
Freud, Anna: 216-17, 224, 229
Freud, Sigmund: 15, 16, 17, 23, 27, 28, 37, 38-39, 40, 44, 46, 47, 49, 55, 61, 63, 64, 76, 93, 94, 167, 172-73, 176, 179, 182, 190, 212, 217-20, 221, 225, 241, 242-43
Fried, Michael: 138

Frith, Chris: 226
Gabbard, Krin: 156
Galatea: 134
Gattaca: 145-53, 158, 160, 168, 208, 245
Gautier, Théophile: 120, 121-25, 126, 136, 201, 245
Gender: 5, 36, 37, 39, 42, 55-77, 94, 104, 105, 107, 112, 115, 119, 133, 134, 155, 161, 184-86, 189, 197, 234, 236
Genealogy: 55, 56, 77, 108
Genet, Jean: 107
Gerrard, Nicci: 104
Gide, André: 5, 32, 79-100, 101, 133, 134, 143, 203, 214-15, 244
Gide, Juliette (Rondeaux): 86, 87
Gide, Madeleine (Rondeaux): 90, 96-99, 133, 214-15, 218
Gilbert, Brian: 152
Gilman, Sander: 91
Girard, René: 159
God: 59, 109, 129, 132, 141, 175, 185-86, 224, 243
Godley, Wynne: 241
Goethe, Johann Wolfgang von: 40, 85, 160-61
Gold, J.: 1
Gold, Lauren: 170
Goncourt, Jules de: 121
Gooldin, Sigal: 244
Graciansky, Pierre de: 17
Granoff, Wladimir: 22
Grant, Linda: 104, 112

Grasp/haptic: 45, 129, 181, 186, 198, 204, 205, 207-08
Grosz, Elizabeth: 94, 224, 226, 242
Groundhog Day: 150
Group dynamics: 5, 20, 23, 33-37, 45, 63-67, 75, 242
Group psychology and the analysis of the ego: 37
Guattari Félix: 179, 208, 242
Guntram: 109
Hall, Stuart: 112
Hamilton, Alan: 118
Hand: 32, 47, 49, 75, 88, 91, 112, 163, 165, 176, 190, 202, 203-08, 227
Hardyment, Christina: 148
Harlow, Harry: 45,
Harrison, Kathryn: 229
Hartmann, Heinz: 20
Hatoum, Mona: 169-70
Hauser, Kaspar: 3
Hawke, Ethan: 145
Heller Morton A.: 3
Heloïse: 161
Henry II: 109
Herbart, Pierre: 88-89, 143
Herder, Johann Gottfried: 136, 245
Héritier-Augé, Françoise: 94
Hermaphrodite: 123, 245
Hesse, Eva: 137
Heterosexuality: 58, 61, 72, 79, 136, 160, 190, 214, 222, 245

Highsmith, Patricia: 103, 154-59, 245
Hoffmann, E. T. A.: 130, 134
Hofmannsthal, Hugo von: 127
Hoggart, Simon: 103, 244
Homosexuality: 71, 79, 95, 106, 136, 155, 156, 187, 190, 214
Honigsbaum, Mark: 115
Horse: 53, 57, 132, 134, 173
Horsnell, Michael: 244
Hovering: 171, 174-79, 189, 222
Howes David: 3
Hudson, Brenda: 244
Hugo, Victor: 172, 175, 212-14
Hunger: 91, 92, 96, 98, 99, 112, 186
Hurd, Lord Douglas: 103
Hydraulics: 91, 93-94, 111, 150
Hysteria/hysteric: 48, 103-04, 190
Ideal ego: 37, 64, 66, 242
Imaginary friend: 116, 221-23, 224, 225,
Intelligence: 6, 56, 57, 63, 64, 75, 195, 196, 198
Internal body: 4, 60, 68, 140, 169-70, 171, 180, 196, 212, 213-21, 242, 245
IPA (International Psychoanalytic Association): 21
Irigaray, Luce: 180, 185, 187-89, 197, 211, 246
Itching: 45, 79-80, 83-85, 90, 99-100, 169
Jadin, Jean-Marie: 95

James, Henry: 28, 161
Jay, Martin: 178
Jenkins, Simon: 103, 110, 111, 244
Jesus: 49, 115, 221-22
Jobling, Ray: 84, 85,
Johnson, Boris: 108, 244
Jonah: 180
Jonze, Spike: 145, 166, 245
Josipovici, Gabriel: 1, 4, 208
Judd, Donald: 136-37
Jupiter: 224
Kaës, René: 20, 26, 27, 34, 35, 36, 37, 65
Kafka, Franz: 36, 91-92, 101, 173, 174, 179, 243, 244
Kantorowitz, Barbara: 244
Kapoor, Anish: 139-41, 143,
Katz, Claire Elise: 185
Kaufman, Charlie: 161, 166, 245
Keane, Molly: 143
Keener, Catherine: 162
Keller, Gottfried: 134
Keller, James R: 156
Kennedy, John F.: 104
Kernel: 37, 43, 47, 50, 51, 99, 137, 139, 193, 219
Khan Masud: 241
Kipling, Rudyard: 109
Klein, Melanie: 38, 46, 63, 137, 213, 214, 217, 220, 222
Kleist, Heinrich von: 128-29, 161
Knickmeyer, Rebecca: 135
Körner, Hans: 138
Kremer, Michael: 224

Kristeva, Julia: 94
L'Auto-analyse de Freud et la découverte de la psychanalyse: 17, 38-39
L'Épiderme nomade et la peau psychique: 49, 195
L'Être et le néant: 180-82
L'immoraliste: 90, 214
La Confession d'un enfant du siècle: 170
La Démangeaison: 84
La Dynamique des groupes restreints: 24, 33-35
La Femme sans qualité: 56
La Porte étroite: 86, 95, 214
La Symphonie pastorale: 214
La Tentation de Saint Antoine: 144
Lacan, Jacques: 8, 12, 13, 18-22, 28, 38, 44, 65, 71, 95, 179, 241
Lagache, Daniel: 17, 22, 241
Lambert, Jean: 88
Laocoon: 137
Laqueur, Thomas: 55, 71, 72, 94
Larbaud, Valery: 244
Law, Jude: 145, 146, 157, 158, 160, 164, 165, 166
Le Balcon: 107
Le Corps de l'œuvre: 39, 67-72, 75, 90, 173-74, 229
Le Détracteur: 12
Le Diable et le bon Dieu: 110
Le Groupe et l'inconscient: 35-38

Le Moi-peau: 28, 44-49, 59, 199, 242
Le Penser: 52, 67, 193
Le Psychodrame analytique chez l'enfant et l'adolescent: 17, 32
Le Visible et l'invisible: 6, 183-84
Leclaire, Serge: 21, 22
Leconte de Lisle, Charles-Marie-René: 176-77
Leder, Drew: 4, 169
Leppmann, Wolfgang: 233
Leprosy: 84, 85, 109-10
Les Cahiers et les poésies d'André Walter: 214-15, 218
Les Caves du Vatican: 83, 89
Les Contemplations: 212-13
Les Enveloppes psychiques: 31, 49, 102
Les Faux-monnayeurs: 93
Les Pensées: 17, 50
Lessana, Marie-Magdeleine: 8
Lessing, Gotthold Ephraim: 137
Lévinas, Emmanuel: 180, 184-87, 188, 189, 197-98, 199, 201
Leviticus: 85
Levy, Stephen: 116, 221
Lewin, Kurt: 20
Lincoln, Sarah: 105, 109
Lingis, Alphonso: 179, 185
Litch, Mary M: 145
Loss: 5, 41, 45, 64, 179, 212-39
Love and Death in the American Novel: 86

Love: 5, 29, 52, 73, 74, 81, 106, 121, 131, 147, 156, 162, 163, 167, 169-200, 201, 207, 208-12, 217, 224, 228, 230-31
Lyotard, Jean-François: 179
MacDonald, Marianne: 108
Maclachlan Ian: 242
Madame Bovary: 143-44
Mademoiselle de Maupin: 121, 124-25, 245
Madonna: 110
Maisonneuve, Jean: 25
Malcolm, Janet: 55
Malkovich, John: 145, 164, 165, 166, 245
Mallarmé, Stéphane: 122, 176
Marinetti, Filippo Tommaso: 178-79
Marks, Laura U.: 3, 170, 207-08
Mars-Jones, Adam: 109, 116, 221
Marsyas: 139-41, 143
Marsyas: 176, 229
Martin du Gard, Roger: 81-83, 88,
Martin, Jacques-Yves: 33
Mary: 120, 153,
Masculine/masculinity: 13, 29, 36, 39-40, 61, 62, 65, 68, 69-70, 71, 74, 93, 128, 135, 136, 146, 153, 155, 160, 161, 165, 171, 173, 177, 179, 185-86, 187, 191, 206, 235
Maternal: 39, 42, 47, 61, 63, 68-69, 73
Maximes et pensées: 244
May '68: 23-26

Mazis, Glenn A: 183
Mazzio, Carla: 4
McCrum, Robert: 244
McGonigle, Dave: 226
McKnight, Sam: 104
McLuhan, Marshall: 3
Medusa/medusan: 120, 121, 125, 127
Meltzer, Donald: 5, 36,
Melville, Herman: 42
Memento: 145, 227
Merleau-Ponty, Maurice: 6, 180, 182-84, 186, 189, 198, 201
Meyer, Conrad Ferdinand: 120, 121, 125, 126
Meyers, Carol: 186
Milk: 61, 70, 96, 191
Miller, Jacques-Alain: 4, 28, 169
Miller, William Ian: 84, 94, 99
Millot, Catherine: 95
Milne, A. A.: 222
Minghella, Anthony: 145, 153-60, 245
Mirren, Helen: 117
Mitchell, Silas Weir: 223
Moby Dick: 42
Moebius strip/ring: 43, 75, 195
Mon Abécédaire: 28
Mona Lisa: 115
Monckton, Rosa: 110, 111
Monitor: 23, 24, 33, 35-37, 63, 64-67
Monroe, Marilyn: 114
Montagu, Ashley: 3, 4, 45
Moore, Suzanne: 110

Moorjani, Angela: 242
Moreno, Jacob: 32
Morra, Joanne: 223
Morris, Jan: 112
Morris, Robert: 136
Morton, Andrew: 101, 102, 104, 108, 110, 113
Moses and Monotheism: 243
Moses: 120
Mother/mothering: 15, 19, 31, 33, 34, 37, 40, 41, 42, 43, 56, 62, 64, 71, 74, 97, 98, 99, 103, 128, 190, 214, 233, 234-35, 239
Mother's body: 35, 36, 37, 46, 61, 64, 67, 68, 74, 75, 76, 90, 135, 141, 191-92
Mother-child couple: 6, 36, 45, 47, 48, 51, 62, 74, 187, 190, 192-93, 195-97, 203, 205-06, 214, 217, 218, 224, 227, 231-32, 234, 239
Mower, Sarah: 110, 244
Moyaert, Paul: 185
Mucous: 48, 73, 189, 191, 197, 246
Musil, Robert: 128, 242
Musset, Alfred de: 144, 170, 218
Mutual/reciprocal inclusion: 36, 46, 47, 61, 64, 75, 76, 181, 193-94, 197, 198, 201, 230, 246-47
Nancy, Jean-Luc: 49
Nanterre (Paris X): 22-26, 28

Narcissism: 30, 40, 41, 48, 64, 71, 79, 90, 104, 167, 190, 197, 242
Neidich, Warren: 225
Neill, Sam: 206
Nelson, Lord Horatio: 224
Nero: 176
Nerval, Gérard de: 218
Niccol, Andrew: 145, 152, 208, 245
Nietzsche, Friedrich: 119
Nobécourt, Lorette: 84
Now we are six: 222
Nucleus: 128, 187, 196, 219, 220, 229, 239
Oedipus: 30, 34
Ogilvy, Margaret: 231-32
Olivier, Edith: 222-23
Orgasm: 61, 81-83, 90, 190
Orifices: 32, 44, 48, 93, 94
Orpheus: 215-16, 227
Oz, Amos: 241
Panorama Diana interview: 101, 108, 111
Pantaine, Élise: 7, 8, 10, 12, 15, 30, 238
Pantaine, Jeanne: 7-8, 238
Pantaine, Marguerite (death by fire): 7-10, 133, 238
Paranoia: 8, 11, 13,
Paré, Ambroise: 2224
Parinaud, André: 231
Paris: 12, 16, 90, 98, 172
Parker, Charlie: 156
Parot, Françoise: 19, 53, 241

Parry, Idris: 129
Parthenogenesis: 38, 56, 67, 76
Pascal, Blaise: 17, 50, 90
Paska, Roman: 129
Passivity: 1, 2, 5, 6, 46, 61, 69, 74, 164, 167, 181, 184, 194-95, 198-99, 204
Paternal: 39, 47, 68, 70, 71
Paul, Henri: 116, 117
Penetration: 35, 37, 44, 45, 61, 63, 123-24, 143, 144, 147, 148, 169, 173, 180, 190
Persephone/Proserpine: 213-14
Peter Pan and Wendy: 232
Peters, H. F.: 128
Petot, Jean-Michel: 26
Peyrefitte, Alain: 17
Phallus/phallic: 42, 47, 52, 56, 60, 64, 67, 70, 72, 152-53, 163, 167, 174, 177
Phantom limb: 223-26
Phénoménologie de la perception: 182
Pierre-Quint, Léon: 86, 134
Pinocchio: 134
Pleasantville: 209
Pleasure: 6, 42, 45, 48, 50, 57, 59, 60, 61, 64, 73, 76, 80, 81-83, 84, 93, 101, 109, 181, 187, 190
Plein Soleil: 154, 159, 245
Pluto: 213
Potts, Alex: 135-39
Pouch: 41, 69, 73

Power: 24, 29, 36, 41, 46, 67, 69, 70, 97, 107, 132, 161, 218
Pradier, Jean-Jacques: 136
Prater, Donald: 233
Pregnancy/gestation: 63, 68, 90, 102, 156, 164, 165, 167-68, 196, 215-16
Pre-oedipal: 64, 70
Prince Harry: 118
Prince William: 118
Pritzel, Lotte: 130
Prokaryotic cells: 59, 128, 243
Prosser, Jay: 235, 242
Prosthetics: 129, 148, 149, 151, 205, 223, 224
Proust, Marcel: 3, 86, 173, 174, 178, 212, 246
Pryor, Ian: 202
Psoriasis: 84, 85, 149
Psychic envelopes: 5, 14, 36, 40, 49, 50, 193, 198, 219
Psychoanalysis as practice: 33, 38, 49, 62, 63, 72, 75, 190, 192, 201, 218, 220, 222, 236, 239
Psychodrama: 17, 20, 32-33
Pullinger, Kate: 202
Puppets: 114, 128-33, 145, 160-68
Purves, Libby: 112
Pygmalion/pygmalionesque: 120, 125, 127, 134, 136, 201,
Queen Elizabeth II: 113

Radiance/ dazzle/ glow: 102-03, 107, 109, 114, 118, 138, 158-60, 166, 169, 206, 228
Ramachandran Vilayanur: 223, 225, 226-27
Rambaud Henri: 96
Reader: 39
Reay, Vivienne: 117
Rebecca: 237
Reciprocation: 34, 46, 61, 74, 181, 183, 193-94, 198, 216, 219-20, 225
Redburn, Chris: 104
Replacement child: 14, 15, 38, 57, 77, 128, 133, 141, 152, 157, 160, 222, 230-39
Reproduction: 42, 55, 59, 67, 70, 113, 146, 150, 168, 186
Richards, Peter: 84, 85,
Richelle, Marc: 19, 53, 241
Rilke, Phia: 233
Rilke, Rainer Maria: 102, 120, 125-33, 137-39, 174, 176, 201, 203, 208, 222, 232-34, 244-45
Risk: 36, 42, 47, 52, 60, 64, 67, 98, 135, 147, 150, 174, 187, 199, 211-12, 217
Ritchie, Ian: 2
Robinson, Hilary: 246
Rock DJ: 170
Rodaway, Paul: 1, 2, 3, 4, 207
Rodin, Auguste: 126-28, 130, 131, 138, 244
Rogers, Carl: 23
Romanticism: 58, 92

Rondeaux, Lucile: 90
Rorschach tests: 17, 26, 32, 45
Roudinesco, Élisabeth: 18, 19, 20-21, 28
Rowe, 'Paula': 235
Rudd, Les: 110
Ruddick, Sara: 63
Russian dolls: 62, 75, 161, 163, 196, 197, 219
Sabbadini, Andrea: 236
Sacks, Oliver: 1, 224, 226
Saint-Exupéry, Antoine de: 178
Sami-Ali, Mahmoud: 197, 246-47
Sandler, Anne-Marie: 241
Sartre, Jean-Paul: 110, 169, 175-76, 180-82, 186, 189, 198-200, 204, 208, 244
Scaglia, Hector: 36
Schiff, William: 3
Schilder, Paul: 119
Schlumberger, Jean: 86,
Schnack, Ingeborg: 128
Schwarzenegger, Arnold: 165
Screen: 107, 109, 127, 200, 221
Sculpture/statue: 5, 123-28, 134, 135-41, 184, 186, 198, 244
Séchaud, Évelyne: 189
Second skin: 58, 77, 81, 86, 89, 90, 140-41, 143, 144, 149, 160, 196, 227-29
Sedgwick, Eve Kosofsky: 159, 204
Segal, Leah: 115
Segal, Naomi: 2, 32, 55, 74, 91, 96, 103, 128, 144, 159, 202, 214, 242, 245

Self-analysis: 38, 39, 76
Senses: 1, 6, 44, 47, 68, 184, 201, 203, 243
Serres, Michel: 1, 124
Sex: 44, 48, 52, 55, 58, 65, 67, 73, 79, 80-81, 91, 92, 109, 137-38, 143, 147, 149, 161, 179, 180-200, 206, 207, 229, 243, 246
Sexual difference: 72, 187-89
Sexual indeterminacy: 39, 63, 64, 68, 70, 72, 123-25, 155-56, 164, 182, 228, 234
Sexuality: 59, 67, 69, 84, 93, 96, 98, 99, 103, 134, 136, 139, 155, 180-200, 206, 241-42
SFP (Société Française de Psychanalyse): 20-22
Shakespeare, William: 165
Sheen, Charlie: 165, 168
Sheen, Michael: 117
Shell: 43, 44, 46, 47, 51, 83, 86, 98, 114, 130, 149, 188, 193, 202, 206, 211, 218, 219
Shildrick, Margrit: 94
Showalter, Elaine: 112
Si le Grain ne meurt...: 80, 86, 95, 96-98
Silent Witness: 148
Skin: 4, 27, 35, 37, 42, 43, 44-47, 50, 52, 56, 58, 72, 79, 83, 90, 92, 93, 96, 99, 101, 105, 108, 110, 114, 122, 138, 140, 144, 146, 149, 158, 171, 179, 184, 188, 189, 196, 206, 207, 227, 228, 239, 246
Skin-ego/*moi-peau*: 5, 17, 27, 32, 41, 43, 44, 46-49, 52, 72, 73, 75, 84, 93, 94, 103, 141, 189, 190, 195, 196-97, 201, 208, 212, 242
Smith, Marquard: 223
Sontag, Susan: 85
Soylent Green: 214
Spitting Image: 114
Spitz, René: 51
SPP (Société Psychanalytique de France): 20-22
Stacey, Jackie: 242
Starobinski, Jean: 144
Stéphan, Hervé: 116
Stevens, Sir John: 117
Stewart, Susan: 2
Strachey, James: 241, 242
Street, Sarah: 156
Strong, Roy: 112
Structuralism: 22, 23
Studies on hysteria: 49,
Superego: 42, 53, 57, 86, 100,
Surface: 46, 119, 120, 122, 124, 125, 126-27, 136, 140, 146, 147, 158, 159, 166, 170-71, 180, 193, 208, 220-21, 242
Surveiller et punir: 107
Survol: 40, 174-78
Syrotinski, Michael: 242
Taboo on touching: 49, 51, 100, 171, 189-90, 207, 235-36

Take-off/*décollage*: 40, 67, 68, 172-74, 178, 246
Tang, David: 244
Tarrab, Gilbert: 14, 20, 49, 194
Taupin, Bernie: 115
Temperley, Jane: 241
Tennant, Laura: 103, 244
Thatcher, Margaret: 103, 114
The Battle of Tripoli: 178
The Cherry Orchard: 163
The Flaying of Marsyas: 139
The Fly: 150
The Golden Bowl: 161
The Interpretation of Dreams: 39
The Little White Bird: 232
The Love Child: 222
The Man with the broken nose: 126
The Metaphysics: 70
The Nutcracker: 134
The Physics: 189
The Piano: 32, 201-08, 211
The Queen: 117
The Silence of the Lambs: 159
The Talented Mr Ripley: 145, 152, 153-60, 168, 245
The Truman Show: 201, 208-211
The Waste Land: 221-22
Things/objects: 36, 119-41, 153, 156, 158, 159, 209-10, 227
Thorwaldsen, Bertil: 136
Thought/thinking: 32, 46, 47, 49-52, 67, 68, 71, 73, 99, 198, 200
Thurman, Uma: 146

Tolstoy, Leo: 110
Torok, Maria: 46, 217, 218-21, 223, 229, 236
Totem and taboo: 37
Touch: 3-5, 6, 33, 47, 49, 60, 73, 77, 83, 109, 110, 116, 138, 141, 178, 179-200, 201, 207, 208, 209, 211, 227, 228
Tourneur, Zacharie: 17
Touzard, Hubert: 20
Toxicity: 48, 84, 103, 214
Transference: 33, 55
Tuberculosis: 85, 90
Turkle, Sherry: 20, 26
Turner, Chris: 242
Turquet, P. M.: 45
Tustin, Frances: 46, 47, 135
Tyler, Imogen: 56, 168
Ulrichs, Karl-Heinrich: 86
Une Peau pour les pensées: 9, 52, 194-95
Vagina: 57, 72, 139, 190-91
Valéry, Paul: 127
Valverde, Alejandro: 229
Van Rysselberghe, Maria: 80, 87, 98,
Versace, Donatella: 103, 244
Vidal, Gore: 147
Virgil: 215
Voice: 33, 63, 66, 106, 114, 124, 156, 157, 163, 164, 171, 190, 201-02
Vomiting: 43, 48, 68, 101, 111, 113-14
Wafer, Jack: 244

Waking the Dead: 148
Wallfisch, Benjamin: 246
Wambo, Arnaud: 108
War and Peace: 110
Wax, Ruby: 108
Weir, Peter: 201
Weiss, Gail: 119
Wenders, Wim: 222
Whiteread, Rachel: 137
Whitford, Margaret: 246
Whitman, Walt: 80
Wiene, Robert: 134
Wilde: 152
Williams, Robbie: 170
Williamson, Judith: 116, 221
Wilson, A. N.: 110
Winnicott, D. W.: 33, 39, 41, 45, 50, 51, 63, 74, 99, 134, 179, 196, 241
Wober, Mallory: 3
Womb: 15, 36, 63, 64, 73-74, 76, 150, 174, 191-92, 196, 198, 199, 209, 210, 230
Wood, Aylish: 145
Woods, Vicki: 106, 244
Woolf, Virginia: 112
Wrapping: 13-14, 29, 59, 75, 140, 208, 222, 228, 236
Yeats, William Butler: 178
Young, Hugo: 104
Young-Bruehl, Elisabeth: 229
Zooming: 171, 172-74, 178, 179, 189